SPACE SCIENCE, EXPLORATION AND POLICIES

Progress in Dark Matter Research
J. Val Blain (Editor)
2005. ISBN: 1-59454-243-0

Space Science: New Research
Nick S. Maravell (Editor)
2006. ISBN: 1-60021-005-8

Space Policy and Exploration
William N. Callmers (Editor)
2008. ISBN: 978-1-60456-448-8

Space Commercialization and the Development of Space Law from a Chinese Legal Perspective
Yun Zhao
2009. ISBN: 978-1-60692-244-6

Space Commercialization and the Development of Space Law from a Chinese Legal Perspective
Yun Zhao
2009. ISBN: 978-1-60876-874-5 (E-book)

Space Exploration Research
John H. Denis and Paul D. Aldridge (Editors)
2009. ISBN: 978-1-60692-264-4

Space Exploration Research
John H. Denis and Paul D. Aldridge (Editors)
2009. ISBN: 978-1-61668-212-5 (E-book)

Next Generation of Human Space Flight Systems
Alfred T. Chesley (Editor)
2009. ISBN: 978-1-60692-726-7

Nutritional Biochemistry of Space Flight
Scott M. Smith, Sara R. Zwart, Vickie Kloeris and Martina Heer
2009. ISBN: 978-1-60741-641-8

Environmental Satellites: Weather and Environmental Information Systems
Vincent L. Webber (Editor)
2009. ISBN: 978-1-60692-984-1

Pulsars: Theory, Categories and Applications
Alexander D. Morozov (Editor)
2010. ISBN: 978-1-61668-919-3

Dark Energy: Theories, Developments, and Implications
Karl Lefebvre and Raoul Garcia (Editors)
2010. ISBN: 978-1-61668-271-2

Dark Energy: Theories, Developments, and Implications
Karl Lefebvre and Raoul Garcia (Editors)
2010. ISBN: 978-1-61668-889-9 (E-book)

The International Space Station
David E. Jamison (Editor)
2010. ISBN: 978-1-60692-322-1

Global Positioning Systems
Viggo Asphaug and Elias Sørensen (Editors)
2010. ISBN: 978-1-60741-012-6

Future U.S. Space Launch Capabilities
Franz Lojdahl (Editor)
2010. ISBN: 978-1-60741-384-4

Smaller Satellites Operations Near Geostationary Orbit
Matthew T. Erdner
2010. ISBN: 978-1-60741-181-9

Astrobiology: Physical Origin, Biological Evolution and Spatial Distribution
Simon Hegedűs and Jakob Csonka (Editors)
2010. ISBN: 978-1-60741-290-8

Space Tourism Issues
Elias Wikborg (Editor)
2010. ISBN: 978-1-60741-353-0

Black Holes and Galaxy Formation
Adonis D. Wachter and Raphael J. Propst (Editors)
2010. ISBN: 978-1-60741-703-3

Exploring the Final Frontier : Issues, Plans and Funding for NASA
Dillon S. Maguire (Editor)
2010. ISBN: 978-1-60876-080-0

Primordial Space
Bernd Schmeikal
2010. ISBN: 978-1-60876-781-6

Sustaining the Global Positioning System
Earl M. Peabody (Editor)
2010. ISBN: 978-1-60741-006-5

Analytical Methods for the Formation of Dark Matter Haloes in the Universe
Nicos Hiotelis
2010. ISBN: 978-1-60876-473-0

Space Material Sciences
A.I. Feonychev
2010. ISBN: 978-1-61668-236-1

Space Material Sciences
A.I. Feonychev
2010. ISBN: 978-1-61668-477-8 (E-book)

Ultra High Energy Cosmic Rays: A New Tool for Astrophysics Research
Anna Uryson
2010. ISBN: 978-1-61668-847-9

Commercial Space Transportation
Jocelyn S. Gunther (Editor)
2010. ISBN: 978-1-61668-707-6

Commercial Space Transportation
Jocelyn S. Gunther (Editor)
2010. ISBN: 978-1-61668-795-3 (E-book)

Future of U.S. Human Spaceflight: Background and Issues
Derek A. Warren and Bridget D. Conway (Editors)
2010. ISBN: 978-1-61668-774-8

Future of U.S. Human Spaceflight: Background and Issues
Derek A. Warren and Bridget D. Conway (Editors)
2010. ISBN: 978-1-61668-876-9 (E-book)

NASA's Future and it's Pursuits
Sophia C. Correno (Editor)
2010. ISBN: 978-1-61668-999-5

NASA's Future and it's Pursuits
Sophia C. Correno (Editor)
2010. ISBN: 978-1-61728-063-4 (E-book)

SPACE SCIENCE, EXPLORATION AND POLICIES

ASTROBIOLOGY: PHYSICAL ORIGIN, BIOLOGICAL EVOLUTION AND SPATIAL DISTRIBUTION

SIMON HEGEDŰS
AND
JAKOB CSONKA
EDITORS

Nova Science Publishers, Inc.
New York

LIBRARY OF CONGRESS CATALOGING-IN-PUBLICATION DATA

Astrobiology : physical origin, biological evolution, and spatial
distribution / editors, Simon Hegedus and Jakob Csonka.
p. cm.
Includes index.
ISBN 978-1-60741-290-8 (hardcover)
1. Exobiology. 2. Evolution (Biology) I. Hegedus, Simon. II. Csonka, Jakob.
QH326.A88 2009
576.8'39--dc22

2009051571

Published by Nova Science Publishers, Inc. † New York

CONTENTS

Preface		**ix**
Chapter 1	**Learning about Astrobiology: A Challenge for the Public Understanding of Science** *Keith S. Taber*	**1**
Chapter 2	**Water Interfacial Processes in Prebiotic Chemistry** *Marta Ruiz-Bermejo, César Menor-Salván, María-Paz Zorzano, Zoubir El-Hachemi, Susana Osuna-Esteban and Sabino Veintemillas-Verdaguer*	**27**
Chapter 3	**Integration of Primitive Metabolic, Information and Structural Protocell Components under Simulated Early Earth Conditions** *S. E. Maurer and P. A. Monnard*	**59**
Chapter 4	**Microbial Evolution and Adaptation in Icy Worlds** *Alberto Alcazar, Laura Garcia-Descalzo and Cristina Cid*	**81**
Chapter 5	**Seepages and the Astrobiological Potential of Polar Dunes on Mars** *A. Kereszturi, D. Möhlmann, Sz Berczi, T. Ganti, A. Horvath, A. Kuti and A. Sik*	**97**
Chapter 6	**Prehistory of the Initial Microbial Communities: The Way from Abiotic Organic Microsystems to Prokaryotic Cells** *Vladimir Kompanichenko*	**115**
Chapter 7	**Visualization in the Education of Astrobiology** *Akos Kereszturi*	**131**
Chapter 8	**The Search for Life on Mars: Raman Spectroscopy and Analytical Astrobiology** *Howell G. M. Edwards and Michael D. Hargreaves*	**143**

Chapter 9 **Europa: New Horizons for Astrobiology** **193**
Esther Velasco, Guillermo Caravantes, Beatriz Gómez-Izquierdo, Alberto Rodriguez and Javier Ruiz

Index **205**

PREFACE

Astrobiology is the study of the origin, evolution, distribution, and future of life in the universe. This book offers an analysis undertaken from the perspective of science education, to discuss the nature of the particular challenges faced by those seeking to facilitate public education in the field of astrobiology. The authors also address the question of the formation of a protocell and its function at the system level. It will further deal with the environmental pressures likely present on the early Earth, as they would have influenced the lifetime of unstable or reactive components needed for the start of life. In addition, the method of the transition from prebiotic organic microsystems to the primordial communities of prokaryotes is explored. The authors also explain some aspects related to aqueous interfacial processes in the context of prebiotic chemistry under the point of view of our current experimental results. Moreover, there is considerable interest in astrobiology in investigating icy worlds and the microbial forms that thrive in extreme environments, especially under those cold conditions that can provide a model for some extraterrestrial environments. This book reviews the general trends concerning the biodiversity and ecology of psychrophilic microorganisms and their molecular mechanisms of adaptation and evolution, as well as summarize the latest results on this topic in samples from icy environments. Other chapters summarize the morphological signatures of flowing water, the astrobiological potential of the dunes, and the theoretical modelling background of the presence of liquid water on Mars today.

Chapter 1 - Over recent years the importance of the public understanding of science has been increasingly recognised. Science is not only a core area of culture, but is also a major area of investment from public expenditure. Awareness of the potential of science to develop technologies which can safeguard and extend the quality of life - for example, through medical advances - is matched by knowledge of the role science has played in developing increasingly effective mass weapon systems and in facilitating the non-sustainable exploitation of natural resources. The public has a somewhat ambivalent view on whether science is in general a good thing, and scientific careers are not seen as attractive to many talented young people. Scientists in all fields must therefore engage with the issue of 'public relations' rather than consider their work as only the concern of an intellectual elite. Education for public understanding has become one of the responsibilities of those working in research areas funded from the public purse. Furthermore, spending money on aspects of 'space science' may not seem a priority compared with developing renewable energy technologies, tackling pollution and protecting biodiversity here on earth. Astrobiology has some advantages in this regard, being an area that can readily inspire the public imagination. Conversely, the very familiarity with aliens represented in mass media, may also mean that scientists attempting to explain the nature and significance of their work to the public may often face unhelpful preconceptions. Moreover, it can be argued that the areas of science that astrobiology draws upon (such as chemical, biological and cosmological evolution) are among those where research into science education has identified major learning difficulties. The present chapter offers an analysis undertaken from the perspective of science education, to discuss the nature of the particular challenges faced by those seeking to facilitate public education in the field of astrobiology.

Chapter 2 - Prebiotic chemistry studies the possible origins of the building blocks of living organisms, as well as their chirality. The hypothesis commonly accepted to explain the origin of life is that simple inorganic molecules, for instance, CH_4, CO_2, N_2, NH_3, reacted together in order to form simple organic compounds that then underwent subsequent reactions to yield biologically functional molecules. These compounds self-organized and self-assembled to increase the organization at molecular levels. This succession of chemical processes is considered the most likely pathway for the emergence of the first living organism. Therefore, the first step to verify this theory is to check how the components of proteins and nucleic acids, among others, could be formed under abiotic conditions. Additionally, the chirality of the constituents of these polymers is another important point. A chiral molecule is a molecule that is not superimposable to its mirror image, L and D configuration. All amino acids in nature have L configuration and all sugars are D. To date, the origin of homochirality in nature is still not understood.

The presence of water is considered a fundamental factor for the development of life. Approximately 72% of the surface of the Earth is covered by liquid water in the form of oceans. This huge water pool and the atmosphere are connected by a cycle of evaporation, condensation, and rain. In addition, oceanic organic and inorganic surface products can be transmitted within aerosols to the atmosphere above, where they are then exposed to radiation and electrical discharges in the form of lightning. The terrestrial water cycle and its changes of state and distribution must have played a significant role in the origin of organic molecules under abiotic conditions, as well as in the symmetry breaking.

In this review, we explain some aspects related to aqueous interfacial processes in the context of prebiotic chemistry under the point of view of our current experimental results. We propose a possible role for the cycle of natural aerosols (i.e., from the ocean water pool to the atmosphere and back to the ocean, bubble-aerosol-droplet cycle) and demonstrate the importance of melting-freeze water cycles. These processes may play a relevant role in the synthesis of organic molecules directly related to the first self-organized systems. We also explain a plausible role for the nucleation processes in the solid-vapor interface in the symmetry breaking.

Chapter 3 - The first forms of living system, or protocells, must have been composed of chemical components or subsystems allowing them to ensure their maintenance, growth, and evolution. As such these subsystems must have possessed similar functions to those of present cellular life. However, contemporary cells rely on the interconnected work of information polymers, catalytic polymers, and their compartments, whose design required a long evolution.

To model plausible protocells and their functions, several classes of simpler compounds have been proposed and investigated, such as single chain amphiphiles (e.g., fatty acids) as building blocks of the compartments, polycyclic aromatic hydrocarbons (PAHs) as energy harvesting molecules, and catalytic RNAs (ribozymes) as both information and catalytic components. While very interesting insights in the processes and/or functions of each molecule class have already been gathered, it has become increasingly clear that the study of the protocell as a whole must be carried out to achieve the construction of a truly functional model system.

This review addresses the question of the formation of a protocell and its function at the system level. It will further deal with the environmental pressures likely present on the early Earth, as they would have influenced the lifetime of reactive components needed for the start of life. The capabilities and limitations of protocell functionalities can be elucidated only through understanding the container interactions/interconnected functions with the other subsystems including environmental factors.

Chapter 4 - There is considerable interest in astrobiology in investigating icy worlds and the microbial forms that thrive in extreme environments, especially under those cold conditions that can provide a model for some extraterrestrial environments. After the successful missions Galileo and Cassini-Huygens to Jupiter and Saturn systems respectively this priority has been focused on the research of environments potentially habitable in some icy worlds such as Europa, Encelado or Titan.

On Earth, the recent discovery of cold-tolerant microorganisms in glaciated and permanently frozen environments has broadened the known range of environmental conditions which support microbial life. These microorganisms, known as psychrophiles are defined as organisms having an optimal temperature for growth at about 15ºC or lower, a maximal temperature for growth at about 20ºC, and a minimal temperature for growth at 0ºC or below. The lowest temperature limit for life seems to be around -20ºC, which is the value reported for bacteria living in permafrost soil and in sea ice. Terrestrial models of extraterrestrial icy worlds are being intensively studied. Among them, bacterial communities from Arctic and Antarctic permafrost, subglacial lakes and high mountains are considered representative of these environments in which psychrophile bacteria are the unique inhabitants. When compared to other known prokaryotes, psychrophilic bacteria posses many unique qualities and molecular mechanisms that allow their adaptation to cold environments. In this regard, some psychrophiles have been found to contain polyunsaturated fatty acids in their plasma membranes, which generally do not occur in other prokaryotes. Further, some of them use enzymes that continue to function at near freezing ambient temperatures. And finally, psychrophilic bacteria are able to produce proteins that are stable at cold temperatures such as heat shock proteins (Hsps) and cold shock proteins (Csps). This chapter will review the general trends concerning the biodiversity and ecology of psychrophilic microorganisms and their molecular mechanisms of adaptation and evolution, as well as summarize our latest results on this topic in samples from icy environments.

Chapter 5 - At the polar dunes of Mars branching dark slope streaks emanating from Dark Dune Spots were observed in local springtime on MOC and HiRISE images. They seep downward with an average speed of 1.4 meter/sol when the temperature of the terrain on 3 km spatial resolution is between 180 and 250 K. Theoretical considerations and calculations suggest interfacial liquid-like water has to be present on Mars, and possibly is causing the observed seepage-like features. The thin water lubricated grains may slide over each other, and/or the stress caused by the refreezing of this interfacial water could also produce such markings, as well as the moving front of solid ice / liquid water phase changes. Such an ephemerally wet environment between the dune grains may have astrobiological significance.

Chapter 6 - This chapter is devoted to consideration of the method of the transition from prebiotic organic microsystems to the primordial communities of prokaryotes. During the initial transformations prebiotic organic microsystems acquired the unstable bifurcate state, with its following relative stabilization by means of balanced oscillations around the bifurcation point. Then microsystems of the type converted into the primary living units – probionts, which were characterized with the ability to concentrate free energy and information already. Through the synergy effect probionts were organized into the communities that represented themselves minimum self-sufficient unit of life. The appeared communities of probionts evolved to more complex communities of progenotes (characterized by C.Woese) and then prokaryotes. This earliest biological evolution proceeded through development of the initial nucleoprotein interaction, formation of RNA and then DNA, etc. Emergence of species had led to conversion of the initial communities of probionts and progenotes ("embryos" of ecosystems) into the real ecological systems, which were hierarchically organized of different microbial species.

Chapter 7 - This chapter summarizes the experiences regarding the general knowledge of undergraduate students in astrobiology, and presents methods we used in the education of astrobiology at university level in Hungary. Based on these, we found the following important issues. They have good general knowledge in the fields, which are well known for decades. Liquid water is thought to be widespread by the students, but they lack important information (like oceans inside icy moons) in the recent results. The most important is to develop and improve methods for the students to use the knowledge learned at different subjects more frequent in astrobiology. Useful examples, like the astrobiology- and Drake-matrixes are presented to compare different environments based on numerical values, graphics and tables.

Chapter 8 - The analytical application of Raman spectroscopy to the identification of the survival strategies of terrestrial extremophiles through the characterisation of key biomolecular and biological signatures has provided novel information about the roles of protective chemicals synthesised by the organisms under environmentally stressed conditions. The non-destructive sampling requirement and the ability to discriminate between organic biomolecular spectral signatures and those from the inorganic host matrix in the geological record are essential requirements for the assessment of the survival strategies being employed against low wavelength ultraviolet radiation insolation, desiccation and extremes of temperatures found in terrestrial habitats termed Mars analogue sites (Table 1).

The recent announcement of the inclusion of a miniaturised Raman spectrometer as part of a combined Raman /LIBS instrument for the molecular and elemental analytical characterisation of specimens in the Martian surface and subsurface as a component of the European Space Agency ExoMars life-detection suite has focussed attention on the Raman spectra of key biomarkers of life signatures. To search for indicators of extinct or extant life signatures on Mars serves to highlight the importance of Raman spectroscopic techniques in this astrobiological mission. As a forerunner to the ExoMars mission it is imperative that the Raman spectroscopic data from extremophilic colonies from as wide a possible range of terrestrial sites of relevance to Mars is undertaken. This chapter will address the type of information that is required and which is accessible from these terrestrial studies to adequately prepare the way for the interpretation of analytical scientific data that will emerge from the Mars exploration programme, commencing with the launch of ExoMars in 2013.

Chapter 9 - Europa is the smallest of the Galilean satellites of Jupiter, although it is however a remarkable body and the primary focus of astrobiological interest in the Jovian system. Geological and magnetic evidences, as well as theoretical considerations, strongly suggest that Europa has a thin ice shell, maybe a few kilometers or a few tens of kilometers thick, which floats on an internal ocean of liquid water. Some anomalies have been observed in the huge magnetic field of Jupiter whenever Europa goes through it, which seems to indicate that this ocean is made of conductive water, due to a probable high content of salts. Moreover,the low density of impact craters suggest an surface age of about 60 Myr in average, which is very young in geological terms, indicating that internal activity, driven by tidal heating, is resurfacing the moon at the present time. So, Europa shows an overprint of crisscrossing ridges and linear markings, some extending for thousand of kilometers, which are locally disrupted for chaotic terrains, building a colorful and exotic textured surface. All of these features make of Europa a theoretically good niche for life. Biological organisms might exist in the salty internal ocean, in the icy shell (active or latent) at places such as fractures, veins or liquid reservoirs, on the internal sea floor in hydrothermal vents or submarine volcanoes, or even within the rocky interior.

In: Astrobiology: Physical Origin, Biological Evolution... ISBN: 978-1-60741-290-8
Editors: S. Hegedus, J. Csonka, pp. 1-25

Chapter 1

LEARNING ABOUT ASTROBIOLOGY: A CHALLENGE FOR THE PUBLIC UNDERSTANDING OF SCIENCE

Keith S. Taber*
Faculty of Education, University of Cambridge, UK

ABSTRACT

Over recent years the importance of the public understanding of science has been increasingly recognised. Science is not only a core area of culture, but is also a major area of investment from public expenditure. Awareness of the potential of science to develop technologies which can safeguard and extend the quality of life - for example, through medical advances - is matched by knowledge of the role science has played in developing increasingly effective mass weapon systems and in facilitating the non-sustainable exploitation of natural resources. The public has a somewhat ambivalent view on whether science is in general a good thing, and scientific careers are not seen as attractive to many talented young people. Scientists in all fields must therefore engage with the issue of 'public relations' rather than consider their work as only the concern of an intellectual elite. Education for public understanding has become one of the responsibilities of those working in research areas funded from the public purse. Furthermore, spending money on aspects of 'space science' may not seem a priority compared with developing renewable energy technologies, tackling pollution and protecting biodiversity here on earth. Astrobiology has some advantages in this regard, being an area that can readily inspire the public imagination. Conversely, the very familiarity with aliens represented in mass media, may also mean that scientists attempting to explain the nature and significance of their work to the public may often face unhelpful preconceptions. Moreover, it can be argued that the areas of science that astrobiology draws upon (such as chemical, biological and cosmological evolution) are among those where research into science education has identified major learning difficulties. The present chapter offers an analysis undertaken from the perspective of science education, to discuss the nature of the

* Corresponding author: University Senior Lecturer in Science Education, Science Education Centre, University of Cambridge Faculty of Education, 184 Hills Road, Cambridge CB2 8PQ, United Kingdom http://www.educ.cam.ac.uk/staff/taber.html, http://people.pwf.cam.ac.uk/kst24/ E-mail: kst24@cam.ac.uk

particular challenges faced by those seeking to facilitate public education in the field of astrobiology.

Keywords: astrobiology; science communication; public understanding of science; learning science; alternative conceptions; understanding evolution; constructivism

INTRODUCTION

When the novelist and scientist C P Snow wrote about the 'two cultures' he drew attention to how even some of the most erudite intellectuals could claim ignorance of science almost as a badge of honour (Snow, 1959/1998). Whilst all well-educated people are expected to have some knowledge of Shakespeare, Picasso and Mozart, it is quite acceptable to not recognise names such as Bohr, Heisenberg or Dirac. Moreover, whereas being able to distinguish the music of Beethoven from that of Bach could be considered a fundamental prerequisite to being judged cultured, understanding something of Einstein's work would commonly be considered the preserve of 'boffins'.

Despite this, the importance of science education has long been recognised. For example, in the UK, debates about how science could best be learnt in the schools date to the late nineteenth century (Jenkins, 1979). The launching of Sputnik by the CCCP (USSR) brought into focus the need for effective science education in the US, seen as essential to provide the scientific and technological workforce necessary for a country's economic success. This is certainly a lesson that has been taken very seriously by some of the world's fastest developing nations, and remains a strong motivation for prioritizing 'STEM' (Science, Technology, Engineering, Mathematics) education around the world (Jenkins, 1997).

The Importance of Scientific Literacy

However, in recent years, there has been growing awareness that the importance of science education does not solely rest on its utility for the minority of pupils who will go on to study science at higher levels and take up scientific or technical careers. Science education must meet the needs of *all* learners, and that has increasingly been related to developing 'scientific literacy' (Millar & Osborne, 1998). The gist of the argument is that we *all* live in a world where science and technology have major impacts, and therefore we all need to be able to engage with science with confidence and an awareness of both basic scientific ideas and something of the nature of what science is (Taber, 2006).

At a mundane level, as consumers, we are inundated with products claiming technological edge. The critical consumer has to ask, *inter alia*,

- Is there genuine added value in having protein in our shampoo, or five blades on our razor?
- Can more expensive washing powders wash our clothes whiter-than-white; and do the added enzymes really 'digest' stains?

- Will magnets in our pockets clean our blood; or copper bracelets ward off rheumatism; or crystals hung around our necks focus some form of spiritual energy?
- What is the evidence that homeopathy can cure disease, and are there any explanatory mechanisms for how it could work that are consistent with our current understanding of anatomy, physiology and biochemistry?

On a more collective scale, governments have to make decisions informed by scientific knowledge: whether to build nuclear power stations; whether to allow 'genetically modified' (GM) produce onto the market; when to close schools to avoid the spread of pandemics, etc. Whilst some of these decisions will rely heavily on recommendations from scientific advisors, in a democracy such decisions have to be explained and justified to the public. If the public does not want nuclear power or GM ingredients in their foodstuffs, they can express their views – ultimately through the ballot box, and so remove governments that make unpopular decisions. A scientifically literate electorate is more likely to be able to understand and weigh-up the arguments made in such debates.

This may be especially important when things go wrong. The Three Mile Island accident has been used to suggest nuclear power is not safe. Popular accounts disagree on how close that incident came to becoming a major disaster. The Irish government has objected to radioactive material detected in their coastal waters, originating in discharges from a nuclear reprocessing plant on the Western coast of England. The public has to decide whether the level of radioactivity detected is a sign that the English facility is releasing dangerous amounts of material, or just evidence for the excellent sensitivity of modern detectors. Even the Chernobyl 'disaster' can be interpreted in very different ways: showing that nuclear power is inherently unsafe, or demonstrating that even a major nuclear accident is fundamentally no worse than other major industrial mishaps – such as at Seveso in Italy or Bhopal in India. And how do members of the public decide if it is ethical to continue to produce nuclear waste that will be 'hot' for many thousands of years, when there seems to be limited consensus on the degree to which we have already identified safe and secure methods of long-term storage? Such issues can only be understood by those with some scientific knowledge and understanding – but the vote of a citizen who is scientifically literate carries no more weight than the vote of one who is largely ignorant of science and technology.

In recent years it has become increasingly clear that there is more at stake here than just industrial progress and economic success. We are told that the planet's biological diversity is being lost at a vastly increased rate, and that this is due to human actions. Many of the believed-to-be-lost species have never even been characterised, and so have never been economically important to humans. Without some understanding of ecology, the arguments for the inherent importance of diversity may seem subtle for people who cannot see how the extinct species could have been harvested and processed to meet their immediate needs.

Forests are disappearing, and deserts are growing – but some argue that such change is 'natural' (by which they mean independent of human action) and inevitable. Climate change is now widely accepted in scientific circles, but again the arguments about the evidence for change, and the possible consequences are complex. If one politician warns of impending disaster, but another rejects global warming, then the public is likely to be confused. Such a scenario was played large in US politics where the election of a President who did not accept the scientific case for climate change 'hung' on judgements about whether tiny hanging chads of card represented intended votes or incidental damage to voting papers. Whereas the man deemed elected to the highest office in the most powerful nation delayed action on greenhouse emissions for years, his presidential rival went on to be awarded a Nobel prize for his efforts to make the case for action. There were of course many substantive issues separating the two candidates, but the close and disputed nature of that election highlights just what is at stake in having an electorate that is well-informed on scientific issues.

It is important to reiterate that scientific literacy is not just about understanding some science, but just as much about understanding the nature of science itself – how it 'works' (Millar & Osborne, 1998; Taber, 2006). When scientists cannot agree on how quickly climate change is taking place, or on how conditions will change in different parts of the world; and when we cannot know for sure whether the process may soon reach another local 'attractor' in the ecosystem (to leave surface conditions stable but slightly different) rather than slip into a runaway greenhouse (that could turn Earth into a twin of Venus), then the casual observer might question how certain any of the relevant science is.

Scientists are skeptical by training and profession - the only thing we really know is that we can not be absolutely sure of anything we know – and so a lack of absolute certainty is part of the familiar intellectual landscape for anyone working in science. For the lay person, technical debates about 'how much' and 'how quickly' may seem a good reason to delay until the scientists can actually agree that they really know what is going on. However, this is one case where waiting for definitive empirical data is not an option: after the 'experiment', the findings may prove to be of purely academic interest if we are left with an uninhabitable world. Arguably, public levels of scientific literacy could here be crucial to decision-making that might determine whether Earth remains hospitable to humans and many other species over the coming decades and centuries. (And so whether Earth remains of interest to potential astrobiologists from elsewhere as a source of collaborators - or simply as an archaeological field site!)

The Public Sponsorship of Science

A major consideration is that much science is publicly funded. Decisions have to be made about which science projects to support from finite public resources. When US President J F Kennedy announced that man would go to the moon he took up a challenge that was widely supported: certainly within the US, where the notion of a 'space race' with the Soviets helped maintain support. The importance of metaphor (Lakoff & Johnson, 1980) is such that common US public perceptions of the Soviets as an 'enemy' (and their Communist comrades in China as a potential 'peril') was a strong motivator for supporting international *competition*. In a post 'cold-war' World it is questionable whether such a priority would have received similar levels of strong support. The greater degree of international *cooperation* in today's space programmes is certainly to be welcomed; but many question prioritising large sums of public money for space science whilst much of the World's population faces unnecessary starvation or early death from preventable disease. The potential of findings to inform decision making about issues vital to living on earth may not be immediately obvious.

Perhaps walking on the moon, another world, offered a particularly appealing target in a way that maintaining an international space station, just on our planetary 'door-step' by comparison, or sending unmanned probes out into the solar system, does not. There is a complex issue to be untangled here. The science of the international space station, or even of the Hubble telescope, is not so readily sold. Putting people into space may even seem foolhardy. The loss of Apollo 1 may be presented as heroic, even if the danger of mixing electrical switches and an oxygen-rich atmosphere seems obvious in retrospect: the losses of Challenger and then Columbia tend to be seen more as a matter of casual incompetence.

THE 'PUBLIC UNDERSTANDING OF SCIENCE' AS AN EMERGING FIELD

For most people, formal science education ends when they leave school. Although a proportion of school-leavers continues to study science in post-compulsory education, most people do not attend formal science lessons in their adult years. This makes school science extremely important: but there is much evidence that for many adults who have passed through the school system, their understanding of science leaves much to be desired.

So there have been many surveys claiming to show that substantial proportions of the populations of developed nations (such as the US, and the UK) fail to appreciate the kinds of basic scientific facts and principles that Lord Snow might have argued should be part of the common heritage of any cultured person.

Moreover, significant proportions of some populations give as much credence to superstitions such as astrology as to science, and many do not accept such a well-established scientific principle as the evolution of life on earth. So Oliveira (2008: 24) reported in 2008 that "in a Republican Debate, three candidates to the Presidency of the United States of America stated that they do not believe in evolution". That is clearly an issue for scientists who wish to disseminate their findings to the public if they are working in astrobiology – a field which is "generally seen to be a science concerned with 'the origin and evolution of life in the Universe' or derivations of this theme" (Cockell, 2002: 263) and which "studies life in the universe, its origins, evolution and distribution" (Rodrigues & Carrapiço, 2006: R-1).

Indeed, there is now even a significant minority in some countries who do not believe that the Apollo programme actually got people to the moon. It has been suggested that "as many as 20% of the US population believe that the Moon landings did not take place but were faked by NASA in a television studio on Earth" (Jones, 2004: 47). Even if (as one might hope) this figure is unreliably high, this still seems a worrying finding from a population that is so easy sold a wide range of incredible ideas and dubiously enhanced products.

Scientists themselves increasingly recognise the importance of informing the public about their work, citing a range of reasons including "public accountability; a better informed public; generating support (financial, social, political) for specific areas of science and engineering; [and] recruitment of students" (The Royal Society, 2005: iv).

In recent years, then, there has developed an active field of *the public understanding of science* (PUS), which has evolved somewhat independently of traditional science education. Science education is more firmly established as an academic field (Fensham, 2004), but has traditionally been concerned largely (if not exclusively) with science learning and teaching in formal contexts (Taber, 2009a). The field of PUS, however, is commonly associated with what is termed 'science *communication*' rather than science *education*. It will be suggested below that this could be a significant hindrance if the PUS movement wants to change people's minds.

PUS has also been associated with the notion of the 'science wars', i.e. taking the side of science against influences in society that are seen as being 'anti-science'. A major difficulty here is that there is no clear consensus on whom the 'enemies of reason' (a term used for example by the inaugural holder of Oxford University's Chair in PUS, Prof. Richard Dawkins) are, or more fundamentally even, on what science actually is.

So religion is seen as an archenemy of science by some commentators: but clearly not by those many active scientists from different religious backgrounds who see no contradiction between their science and their faith (Barbour, 2002). Presumably many of those who practice astrology, Feng Shui, clairvoyance, voodoo and a wide range of other practices that most scientists would frown upon, consider themselves to be working in an established tradition developing a practical craft and/or system of knowledge through a combination of empirical investigations and received wisdom. That is, whilst those in the science camp may claim these practices are neither systematic, nor based on critical examination of evidence: they may be considered by their genuine supporters to meet both these criteria. As Carl Sagan (1995: 10) commented: "People are not stupid. They believe things for reasons. Let us not dismiss pseudoscience or even superstition with contempt". Scientists commonly give great weight to prior authority, and regularly find ways to explain away inconvenient findings, and to an observer standing outside of the scientific community (and so lacking a detailed grasp on the issues), the grounds for doing so may seem no better justified than the arguments of a astrologer or psychic medium.

Marxism was meant to be a scientific approach to understanding socio-historical issues, yet is unlikely to be considered part of science by most professional scientists today. Freud certainly saw himself as working in a scientific tradition (Claxton, 2005), yet today his methods and conclusions are considered by many to be highly subjective.

Are either of these areas science? Freud's work is both widely criticised, and acts as the source of many valuable insights still informing psychiatrists today. Marxism's scientific credentials must for ever be sullied by their application under Stalin, and in particular, the Lysenko affair, where the adherence to Marxist doctrine took precedence over scientific values (Frolov, 1991): failure to defer to the desired conclusions meant career suicide or even imprisonment for scientists, whilst subjugating scientific norms to the metaphysics of Marxism led to starvation for a great many in the wider population. Yet one of the thinkers most influential in education today, Lev Vygotsky (1978), undertook his work committed to and informed by Marxist principles. This is all the more notable as his own work was eventually censured by the State, and only become known and influential in the West decades later, long after his premature death. Others working in the Soviet system, such as Luria (1976; 1987) and the activity theorists (Engeström, Miettinen, & Punamäki, 1999), produced work which is still of great interest day. To decide whether Freudian psychoanalysis or Marxism can be considered part of science, we would need an agreed *demarcation* criterion. However, whilst there have been suggestions on what such a criterion should be, there is no general agreement and probably never could be.

Deciding what is or is not a part of science would surely have to be a consensual view of the scientific community: yet deciding who counts as part of that community depends upon the criterion chosen. In practice, there *is* widespread general agreement on who are scientists and what is science: but with a good deal of fuzziness around the edges. Science is perhaps understood as one of those concepts where it is easy to suggest exemplars or prototypes, but impossible to provide a definitive sharp boundary. The philosophers of science may individually be informative here, but are collectively unhelpful. Popperians are likely to exclude many activities admitted by followers of Kuhn or Lakatos, whereas Feyerabend would mischievously put a case for including many of those practices that are commonly seen as anti-science.

What is clear is that some of the common targets of the science side of this public debate are popular among lay people. So Halpern characterises the American public

> "First, a large percentage start everyday by reading their horoscope and believe that it is so often correct that it's as though it was written especially for them; they phone their personal psychic, at a cost that many cannot afford, for advice on matters that range from how to invest their money to whether or not a loved one should be disconnected from life support systems; they spend huge sums of money on a variety of remedies for which there is no evidence that they work or are even safe to take — sometimes with disastrous results"
>
> Halpern, 2000: 22

Halpern goes on to cite a study suggesting that even among college level students, "more than 99 percent expressed belief in at least one of the following: channeling, clairvoyance, precognition, telepathy, psychic surgery, psychic healing, healing crystals, psychokinesis, astral travel, levitation, the Bermuda triangle mystery, unidentified flying objects (UFOs), plant consciousness, auras, and ghosts". It should be noted this 99% figure did not specifically include formal religious beliefs, although for some taking the 'science' side in the perceived science wars, "belief in ESP, belief in God, and belief in the unluckiness of the number 13 all are tarred with the same brush" (Irwin, 1993: 4).

Astrology supports a community of practitioners who make a living from advising people based (supposedly at least) upon the arrangement of celestial bodies at the date and time of birth. In its most popularist form, even quite respected publications carry 'horoscopes' because many of their customers wish to read them. That these 'horror-scopes' tend to make claims of the sort that about eight percent of a population needs to be careful with money this week, or should look out for the needs of loved ones, or deserve to take some time for themselves, etc, seems laughable to the scientifically-minded. Popular horoscope writers could surely just as easily find a rationale in their system to suggest that the end of the week would be a good time for those with one 'star sign' to carry out that planned criminal act, or that now, whilst their loved ones are distracted, is a good time for infidelity for those of another 'birth sign'. Such advice would be no less based on any real evidence, but would not be so welcome in the popular media. Some scientific greats of the past (such as Kepler) took horoscopes seriously, but at least sought to cast them based on specific details of the individual concerned. However, many otherwise sensible people do not see why some scientists and skeptics get so agitated about this popular form of superstition, seeing it as nothing more than harmless fun.

Horoscopes are just one aspect of a long-standing tradition of using natural signs to interpret human concerns: be it the (effectively random) patterns in tea leaves or animal entrails or interpreting meaning in a sequence of tarot cards or through the use of the I Ching. If the people 'reading' these predictions considered them as a form of 'oblique strategy', then scientists would find them less objectionable. The composer and record producer Brian Eno is well known for using such techniques with musicians as an aid to overcoming creative blocks. If the performers are 'stuck in a rut' then they pick a random card that might suggest they swap instruments or try playing at half tempo or whatever. There is no pretence that the card somehow 'predicts' the right way forward, it just offers a way out of a rut, and introduces some variation that the artists' creative minds can take as a new starting point. If people looked for patterns in tea leaves in a similar expectation – like a do-it-yourself word association or ink-blot test, where the meaning is recognised to come from the interpreter not the stimulus – then these activities might be seen by scientists as no more than self-awareness aids, rather than dishonest appeals to supernatural forces.

The problem with horoscopes is, of course, the belief of many lay people that at some level there is 'something in it' - beyond the ability to focus people's minds on specific aspects of their lives because of their suggestibility.

Learning Science: Constructing Understanding and Responding to Misconceptions

One of the points made earlier in the Chapter was that the PUS movement has tended to be associated with what is commonly described as 'science *communication*' rather than 'science *education*'. Perhaps this is meant as no more than a careful choice of terminology that seeks to avoid alienating the audience. After all, children are sent to school to be educated, where adult citizens seek rather to be informed. The latter are already educated, the argument might go, and science communicators offer information suitable for intelligent, educated adults. Perhaps. Perhaps also there is something of an identity issue here: science educators – it could be argued - are not real scientists, but just teach about other people's work; whereas *real* scientists do the actual science and then communicate it as part of the overall scientific process. Again, perhaps.

In some ways the best people to communicate science might be the scientists themselves: they have the greatest knowledge of, and enthusiasm for, the science; they are in a position to speak with real authority.

These are all pertinent points. However, if the PUS movement is based on a premise that members of the public will change their minds if only they can be told about science first hand by those at the cutting edge, then this may prove a simplistic assumption.

The Challenge of Science Teaching

Teaching is often seen by outsiders as about those with subject expertise informing - through telling, showing and explaining - those lacking that expertise, so that knowledge is 'transferred' from teacher to student. Although the terminology of 'transfer' is widely used, a better term for the process described would be 'copied': as the actual perceived aim is often that the teacher's knowledge should be copied to the minds of the students (Taber, 2009a).

If teaching is seen in these terms then 'science communication' would seem to *substantially* be no different to science teaching: teaching where the learners are often adults and not school children, and the communicator is a genuine subject expert talking about work they are intimately involved in, rather than just a teacher who has usually 'only' learned about the topic secondhand. Science communication could then seem to be something of a higher-grade version of science teaching.

Of course, I am going to suggest that this is hardly the case. It has been known for some decades that carefully and well planned teaching by teachers with good subject knowledge is often insufficient to lead to attentive and motivated pupils acquiring the target knowledge set out in the science curriculum. A range of possible reasons for such failures to learn has been explored.

One approach considered carefully examining the structure of the subject knowledge itself to make sure that the teaching sequence adopted introduced material in the most logical order (Gagné & Briggs, 1974). Another area of research, associated particularly with the work of Jean Piaget (1972), explored the way in which a child's cognitive abilities develop, and suggested that many science topics were not within younger children's grasp (Shayer & Adey, 1981), not at least unless special care was taken to find ways to link them to children's experience (Bruner, 1960). A complementary approach has considered how the brain can be analysed as an information processing system (White, 1998), and has considered the significance of features (such as working memory capacity) that act as system 'bottlenecks' (Miller, 1968).

These different perspectives all offer considerations that do need to be taken into account in effective science teaching (Taber, 2009a). However, they have not proved to be enough. So whilst poorly sequenced, over-paced teaching with a cognitive demand beyond the learners' stages of development is certainly unlikely to bring about intended learning; it does not follow that well-sequenced, well-paced, and well-matched teaching will lead to students acquiring the target knowledge. These considerations are *necessary-but-not-sufficient* for good science teaching – there is something more that needs to be considered (Taber, 2001).

This 'something' is what the learner already 'knows'. Towards the end of the 1970s it became widely accepted that approaches to teaching science based on an implicit assumption that ignorance was to be replaced with knowledge were inadequate. It was found that school children, and even pre-school children, already had a lot of knowledge of many scientific topics (Piaget, 1929/1973). However this was not 'knowledge' in the philosopher's sense of *justified, true belief*: often what the child 'knew' was not well aligned with scientific knowledge, and so was at odds with what they were meant to learn in school – even if from the child's perspective there was good reason for their way of thinking about the topic.

Alternative Conceptions of Science Topics

Often the science teacher was presenting a topic about which the children already had a good many ideas: some quite explicit, others tacit (but none-the-less influential in their thinking); some quite similar to the scientific models, but others that could be in complete contradiction. Clearly some of these ideas had significant potential to interfere with the learning set out in the curriculum.

A few examples of the kinds of student thinking about science topics that have been uncovered in research cannot do justice to the phenomenon, but will at least illustrate the general point. Probably at least four-fifths of students in most introductory physics classes have previously developed a way of thinking about force and motion at odds with Newtonian dynamics (Watts & Zylbersztajn, 1981). Most people's intuition is that a force is needed to keep something moving, and that in the absence of such a force a moving object will come to a stop. Research suggests such ideas are not readily extinguished by teaching, and that even if students dutifully learn the physics model for examinations, they often readily slip back into the 'alternative' conceptual framework when asked questions in a real-life context.

Whilst this example reflects one of the most common and widely explored alternative frameworks, it could be argued that students' ideas are not really wrong here (as in common experience a moving object *will* come to a stop in the absence of some driving force) – but just based on a different formalism (Taber, 2009a). After all, it can be asked, why does it make sense to derive a system that starts from assumptions of no air resistance, gravity or friction, when all familiar examples of objects in motion (footballs, sprinters, cars, planes etc) are found in contexts where frictional forces and gravity operate?

However, a second example where such an argument is less supportable may be useful. Research suggests that by the time students get to study why reactions occur in high school or college chemistry (for example in terms of enthalpy changes), most have already developed their own alternative explanation. For most students chemical reactions occur to allow atoms to fill their shells (Taber, 1998). Again, this is an idea which seems to be resistant to change, and which students will offer even after being taught scientific models. Although this example has been less well explored in different populations than the so-called 'impetus' conception of motion, it is in some ways more telling. For one thing, it cannot be argued that students develop this idea because of the apparent behaviour of familiar examples: all children have extensive experience of objects requiring force to move them long before they formally study physics; but no child has direct experience of the electron movements during reactions to bring to the chemistry class. Even more telling, almost all common chemical reactions, certainly those studied in elementary science and introductory chemistry, are clear counter-examples to the common alternative conceptual framework (that is, reactions occur *despite* the reactant species nearly always already having atoms with full shells). Yet, for example, students asked about the reaction between H_2 and F_2 will commonly 'explain' that the reaction occurs because this allows the hydrogen and fluorine atoms to obtain full shells (Taber, 2002)! When students can commonly hold such ideas in the face of apparently overwhelming counter-evidence, it is clear that effective science teaching can be quite a challenge.

The Nature of Prior Ideas about Science Topics

A vast research programme has explored student thinking across most science topics at different student ages and in many countries (Duit, 2007). This research has used various approaches, but one of the most common and direct techniques is simply to interview children or students and to ask them to explain their ideas (Gilbert, Watts, & Osborne, 1985). To give a flavour of this kind of research, I present later in the Chapter a small extract from an interview with one thirteen year old school girl, Sandra, as she talked about her mental model of space.

Findings from this research programme have been diverse, and indeed sometimes led to considerable debate about what the actual nature of children's scientific thinking is. In retrospect, surveying the range of research undertaken, it now seems clear that children come to science lessons with ideas that vary along a wide range of dimensions (Taber, 2009a).

So whilst sometimes children offer ideas which are little more than romanced suggestions that they will readily drop, on other occasions learners' ideas have been found to be strongly committed to, and to be robust and tenacious enough to be relatively unaffected by teaching. Students' ideas may be similar to scientific ideas, or effectively orthogonal to them, or completely opposed. Sometimes, children's ideas about specific science topics are like conceptual islands with little connection with anything else they seem to know: on other occasions they have developed extensive, theory like networks of propositions into which ideas at odds with school science are well-integrated (Taber, 1998). Sometimes students entertain manifold ideas about a topic and shift their thinking according to the way a question is phrased, or from one example to another (Taber, 2000). However, at other times they have a core way of thinking about a topic that is widely applied across relevant contexts.

It is probably fair to say that children's thinking about science is just a microcosm for human thinking in general. Surely each of us exhibits this range of thinking patterns considering the various conceptual domains in which we operate: including those where we are experts as well as those where we are novices; and those where we have a strong commitment, and those where our interest and motivation are minimal. In view of this wide range of characteristics, it will not be surprising that there does not seem to be a simple prescription for the way that science teachers should respond to students' prior thinking.

Conceptual Change in Science

What this vast body of research has shown, however, is that the 'transfer' metaphor for science teaching is woefully inadequate. Science teaching is not about imprinting ideas on *tabula rasa*, or filling empty vessels with knowledge, it is often about engineering conceptual change.

Learners 'construct' their own knowledge by interpreting new information (whether from direct experience, or reported by others such as teachers) in terms of their existing ideas. This building process operates on quite small 'learning quanta' (Taber, 2005), so – to risk overextending the building metaphor – even if the teacher could make sure the learner was using perfect copies of the teacher's 'concept bricks', they were most unlikely to be used to form a conceptual structure that was a copy of the teacher's own knowledge edifice.

There has therefore been quite an extensive research effort exploring both how conceptual change occurs in science, and how teachers can best engineer it in the classroom. Although there is no reason to assume that *science* education is unique in facing these challenges, as students will bring existing thinking to other school subjects, it has certainly been a strong focus of much conceptual change research (Vosniadou, 2008).

Yet despite the research effort applied to this issue over two decades, and some considerable progress in building useful theoretical perspectives (Chi, 1992; diSessa, 1993; Hammer, 2000; Strike & Posner, 1985; Thagard, 1992), we are still some way from being able to provide teachers with a single general purpose strategy to best shift students' alternative conceptions toward the scientific models, or even a set of strategies within a clear overarching theoretical model showing when to adopt each approach (Taber, 2009a).

What does seem clear is that effective science teaching calls upon at least three major areas of knowledge: subject knowledge of the science to be taught; pedagogic knowledge relating to what is known about the teaching and learning of the topic (for example, what the major learning difficulties are in a topic; which analogies students usually find accessible; which explanations they usually find persuasive); and knowledge of the students themselves. Good teachers engage in perpetual interaction: constantly seeking information about what students already know, what they think, what they understand; judging how teaching is being interpreted; and they use this diagnostic information as the basis for on-line problem-solving to inform the on-going teaching process (Taber, 2002). Needless to say, this is difficult enough in a one-to-one teaching setting, but is highly demanding is classroom contexts.

The Challenge for 'Science Communicators'

Good science teachers are only ever partially successful in getting students to adopt scientific ways of understanding topics, even when they have strong knowledge in all three domains. Professional research scientists, acting as 'science communicators', will often have much deeper understanding of their specialist areas than school or college teachers who have to be familiar with a broad range of subjects and where the target knowledge is not the science itself, but the curriculum models which are designed to match learners at a particular grade level (Taber, 2008b). However, science communicators will often lack the detailed training in science pedagogy, and will often be delivering in non-interactive modes (books, media appearances, talks) where they have minimal knowledge of how what they are saying is being understood and interpreted by their audience. Given the difficulties of communicating scientific ideas faced by professional teachers working interactively with students they usually know well, the challenge for research scientists hoping to bring about conceptual change among an often remote and ill-defined target audience is surely that much greater.

Prior Knowledge and Learning about Astrobiology

What is clear from the educational research literature, is that existing ideas influence new learning. Astrobiology-related ideas have a high profile in the public imagination, as life from other worlds has a well-established place in popular culture. Aliens are highly familiar among fictional characters, including some very well-liked ones: Superman, Mr Spock, ET, Dr Who, etc.: "Aliens abound in our common culture, despite the fact that we currently have no evidence for their existence" (Griffiths, 2004: 179). Although SETI (the search for extraterrestrial intelligence) is certainly part of the wider remit of astrobiology, there is much of great scientific interest that does not require sentient, intelligent aliens - but whilst scientists may get excited about possible traces of micro-organisms elsewhere in the universe, such forms of life are less salient in terms of capturing the public's imagination.

Where there has been a great deal of research into children's and older students' ideas about most school science topics, astrobiology as a field in its own right is an advanced area of study, and has not attracted much direct attention in the science education literature. However, there is much in that literature of relevance to the scientist interested in communicating astrobiology to the public.

Learning about the Earth in Space

One of the fascinating aspects of this area of research is how every individual is likely to offer some new slant on a topic, so that even the experienced interviewer can be surprised at some of the ways young people think. As an example, I will present a small extract of an interview with an English schoolgirl whom I will call Sandra. Sandra was one of a number of students at her school who volunteered to be interviewed as part of a project called the 'Understanding Science Project'[1]. This project was unusual in that rather than focusing on a particular science topic, it was concerned with coherence and progression in student scientific thinking. Students were interviewed over a period of time (several years in the case of a number of students) and simply asked to talk about the science they had been learning in school.

At the time of the interview discussed here, Sandra was nearing the end of her second year of secondary school, so she will have been thirteen years of age. At this stage of her schooling Sandra was studying a topic about earth in space during her science lessons, and I had asked her to tell me about the topic. The extract begins just after Sandra mentions that one can see stars in the night sky.

> Keith: So what's a star?
>
> Sandra: I think it's like a big ball of like, electrically, fire-y sort of thing. ... It's either a big ball of like electric-y stuff, or a big ball of like fire-y stuff. I'm not sure which one. ...
>
> Keith: How can they tell if something is a star or a planet, do you think?
>
> Sandra: I think the planets are generally brighter?
>
> Keith: Why's that?
>
> Sandra: 'cause the sun's rays bounce off them. It's the sun's the only thing, apart from stars, that like has its own light source.
>
> Keith: So what's the light source of the sun?
>
> Sandra: It's like a big ball of fire.

The somewhat vague way of describing things (being like, sort of, thingy) was characteristic of Sandra's talk in these sessions, and seemed to be as much a way her group of friends generally conversed, as an indication of uncertainty.

Researchers interested in learners' ideas have to try to take a neutral stance in terms of not making assumptions beyond what the students tell us. However, researchers - just as much as learners - interpret what we are told through our existing conceptual frameworks, so it is easy to slip up. At this point in the dialogue I assumed that Sandra realised that the sun ('a big ball of like, electrically, fire-y sort of thing') was like any other star ('a big ball of fire'), just nearer.

Keith: Okay, so this is like the big fire-y, electrically thing, yeah?
Sandra: Yeah, except the sun's like really bigger … much bigger.
Keith: So it's much bigger than the stars, is it?
Sandra: Yeah.
Keith: How do you know that?
Sandra: Because you can see it.
Keith: And you can't see the stars?
Sandra: You can see the stars, but you can see that the sun is bigger than the stars.
Keith: Well how big's the sun then?
Sandra: Very big … it's like bigger than earth, it's like bigger than all the planets.

I had not expected this response from Sandra who was a fairly intelligent adolescent with quite a strong interest in science and technology. I was also confident that she understood perspective, and the distinction between something being small, and being a long way away. However, playing devil's advocate I suggested that on the same logic (i.e. that the stars are smaller than the sun because they look smaller) then the sun must be much smaller than the earth,

Keith: Oh that's just silly, because I've seen the sun.
Sandra: But it's like really far away.
Keith: But the earth goes on as far as you can see, and the sun only looks that big [indicating with my hand].
Sandra: Yeah but that's because it is really far away, and so it looks small.

So, as I expected, Sandra was perfectly aware that the larger of two objects will seem smaller when it is much further away. So, surely she could appreciate this possibility with the stars as well?

Keith: So maybe the stars are even further away?
Sandra: No, they're not.
Keith: How do you know that?
Sandra: Because, like, when you go into space, you go past stars.
Keith: When you go where? Into space? [Unsure if I had heard correctly.]
Sandra: When you go into space. [Said as if obvious and commonplace.]
Keith: When have you been going into space?
Sandra: I haven't, but other people have.
Keith: How do you know about this?
Sandra: Because like, it was recorded.
Keith: Okay, and you've seen it on television or something, have you?
Sandra: Yeah.

I know from other interviews that Sandra did watch science and technology documentaries on television, and it seemed she genuinely believed she had seen this. Memory can be very unreliable, and we can often think we remember things accurately when objective evidence suggests otherwise. One interesting study asked students about what they thought was going on in the wires in a simple electrical circuit before, immediately after, and some

[1] See http://people.pwf.cam.ac.uk/kst24/projects/ECLIPSE/USP.html

weeks after a teacher demonstrated how current is conserved round a series circuit. Children commonly tend to expect that the current diminishes around a circuit, being progressively 'used up' at each load (such as a lamp). The study found that even when the teacher's demonstration seemed convincing immediately after the lesson, some time later some of the pupils 'remembered' the demonstration as offering clear evidence for their prior conception that the current will diminish around the circuit (Gauld, 1989).

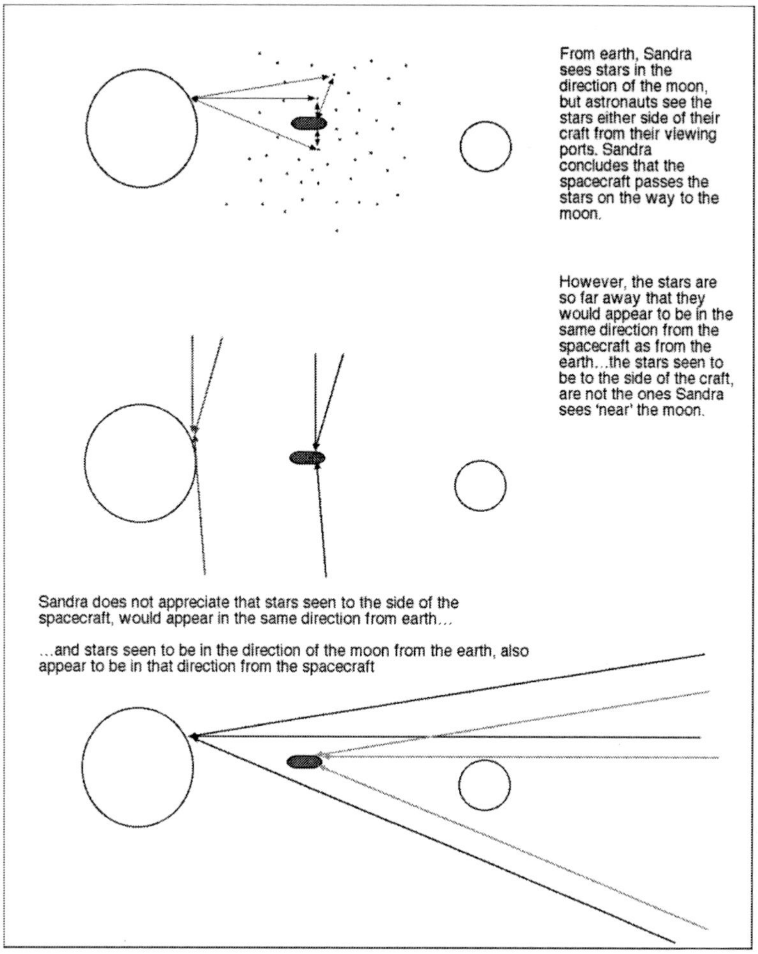

Figure 1. Interpreting Sandra's alternative conception of space

It is not clear what Sandra had seen which led her to form this belief. Perhaps she was confusing documentaries with images from fictional programmes – such as Star Trek where from within spacecraft moving 'at warp' (i.e. at super-luminary speeds supposedly achieved by warping space-time itself) the star field can be seen passing by from viewing ports and screens. Perhaps she saw images of spacecraft such as the shuttle moving against the night sky, and later recalled that the craft was moving past the stars. Perhaps, however, she simply misinterpreted images of 'stationary' stars seen through the windows of real spacecraft according to existing prior knowledge: if there are stars to the left or right of a spacecraft travelling to the moon, and the stars seem to be above us, then surely they must be between us and the moon? (See figure 1). Whatever the origin of this alternative conception, there was nothing wrong with the logic that (based on a false premise regarding relative distances) deduced that stars are smaller than bodies such as the moon or planets.

Whilst this is just one example of a particular alternative conception elicited from one individual schoolchild, there is plenty of research suggesting that misunderstandings about space are common. So, for example, it has been found that youngsters commonly pass through a number of understandings of the relationship between the earth and the universe as they attempt to make sense of what they learn about the earth as a body in space in terms of their common experience of living on an apparently flattish and fixed earth, that people do not seem to fall off of. Students may think that the name earth is also given to one of those bodies out in space (so *that* earth is a round cosmic body like the planets, unlike *this* earth), or that the universe is like a sphere and the earth is the solid hemispherical part of this, with a flat surface on which people live (Nussbaum, 1985).

Despite commonly being taught about the seasons (often a topic met at the top of primary school in the UK), it remains common for students to suggest that the earth is closest to the sun when it is Summer – exactly who's Summer is not usually considered.

There is also a common alternative conception that there is no gravity on the moon *because* there is no air there (with the implication gravity needs air as a medium), whereas the scientific model suggests that the gravitational field strength at the moon's surface is too low to retain an atmosphere. Interestingly, this conception appears to be seen by students as quite consistent with images of people walking, jumping and driving buggies on the moon. The increased span of a step on the moon seems to be seen as some kind of floating, apparently drawing upon an intuitive 'belief' that even without gravity things will eventually ('naturally') end up back on the ground.

It is well recognised that many alternative conceptions found amongst learners reflect, at least superficially, ideas that were once respectable among scientists or their predecessors (Piaget & Garcia, 1989), and it is of note that this particular conception (gravity needs a medium) reflects some of the contemporary concerns about Newton proposing gravity as some kind of occult force acting at a distance.

The most common class of alternative conception about force and motion was mentioned in an earlier section, and these ideas are often labelled 'impetus' thinking and considered to be akin to Aristotle's ideas about dynamics (Gilbert & Zylbersztajn, 1985), albeit lacking the sophistication of Aristotle's own model (Toulmin & Goodfield, 1962/1999). It was suggested above that this conception is found to be highly intuitive, and resistant to teaching.

It can also be the case that even when a Newtonian framework for understanding linear motion is acquired, the physics of circular motion – to which orbital motion commonly approximates – may still be problematic. This can be illustrated in relation to Alice, whom I talked to when she was a student studying university entrance level courses in an English sixth form college. I asked Alive about a range of physics and chemistry topics, looking in particular to explore the level of coherence between her thinking in different topics (Taber, 2008a). Alice had learnt the basic principles inherent in Newton's laws of motion and applied ideas about unbalanced forces being a cause of motion, and specifically acceleration, in a number of situations – falling apples, parachutists etc. She described how increasing air resistance on a parachutist would lead to balanced forces and so a terminal velocity. Although Alice correctly identified unbalanced forces with acceleration, she considered orbital motion to be the result of balancing centripetal and centrifugal forces, that kept the orbiting body moving round. Alice described velocity as speed with a direction, but acceleration as a change *in speed*, and so she did not seem to consider a change of direction alone as a sufficient criterion for acceleration. Although Alice seemed to have a reasonable understanding of balanced and unbalanced forces, and applied this idea across a range of contexts, she misapplied the principle in the context of orbital motion.

Here, in considering circular motion as being the result of balanced forces, she demonstrated another common alternative conception. She thought that the planets were subject to a balance between "centripetal force ... keeping something in an orbit, but also the forces opposing that, which I think is centrifugal...which would send it out of orbit". This seemed to be a misconception of *the status* of circular motion, rather than any difficulty in analysing the situation. In her thinking, orbital motion was *not* accelerated, and so was ontologically similar to uniform linear motion (cf. McCloskey, Carmazza, & Green, 1980).

These brief examples offer just a flavour of the difficulties students find with understanding what we might think of as quite fundamental aspects of space science. Space is huge, such that travel to even the nearest stellar neighbours to our own system would involve very long journeys, even if close-to-light-speed travel becomes possible. Yet in familiar science fiction, moving between worlds inhabited by intelligent beings (usually bipeds, with bilateral symmetry, between 1 and 2 metres high) is commonplace. The universe has evolved over periods of time quite beyond human comprehension in anything other than a formal sense, so that a sense of the relative magnitude of the blip that is humankind is again not something we can readily appreciate. Indeed one study of secondary students' thinking about astronomy concluded that "students' ideas in astronomy are a hybrid of hearsay and imagination, with some tension between the two" (Riga, 2008: 1).

Learning about the Evolution of Living Things

Perhaps at least as challenging as understanding about the physical structure and evolution of the universe, is learning about evolution of its biota. This of course, is a key topic for many of those concerned with the 'science wars' due to the vociferous campaigns by those objecting to the teaching of evolution in schools, or at least claiming that as it is 'just a theory' it should be taught as part of a 'balanced' approach. (In science, an established theory has high status, but those lacking understanding of the nature of science often see a theory as little more than a guess which has not yet been proved.) This usually means also demands for teaching about biblical creation myths as if historical accounts, or at least presenting 'intelligent design' (ID) as an alternative 'theory' on par with natural selection. Rodrigues and Carrapiço (2006: R-4) argue that as "ID advocates explain that the specified complexity that exists in biological organisms can only be explained by being designed by an intelligent agent", its proponents consider "astrobiology a danger for its policies, dogmas and philosophy".

This is a complex and nuanced debate, which may sometimes be obscured by some of the rhetoric that the topic attracts. Certainly among professional biologists, there is very little doubt (a) that evolution did occur, and (b) that the Darwin-Wallace mechanism of natural selection (Darwin & Wallace, 1858) is a key part of the scientific explanation of how evolutions occurs.

As a personal aside, I am writing this Chapter whilst working in the University where Darwin took his degree, at a time when we are especially celebrating his work: 2009 is Cambridge University's 800th anniversary, which coincides with the 200th anniversary of Darwin's own birth, as well as 150 years since the publication of his *On the origin of species* (Darwin, 1859/1968). It seems opportune, then, to point out that whilst it is commonly claimed that Darwin's (and Wallace's) central idea was simple (Dawkins, 1988), this rather underplays the intellectual achievement. Leaving aside his understandable ignorance about what genes might actually be, Darwin's explanation for the origin of life actually requires the coordination of a range of different ideas that have to all be understood, and then integrated into a single explanation, before the principle of natural selection becomes 'simple' (Taber, 2009a). Evolution is one of the most demanding ideas we ask school children to learn, certainly in biology, and probably across science. Indeed, even within biology, there is on-going debate about the precise nature of evolution - not at the level of doubting natural selection takes place, but certainly concerning the precise way it should be understood at the level of genes, organisms, populations and species (Eldredge, 1995).

Consider how some of the participants in the *Understanding Science Project* explained evolution to me. Amy, Jilly and Mandy were nearing the end of their compulsory schooling (at the end of their penultimate year, aged 15), and had been studying the topic for the English school leaving examinations (the General Certificate of Secondary Education):

> "[Evolution is] kind of animals or plants or humans or whatever, like kind of, erm, well, evolving...yeah, to kind of like adapt to their surroundings and stuff....
>
> We were doing this bit on Charles Darwin when he was studying finches in an island somewhere, and he found that like on neighbouring islands that the finches all had different shape beaks because of erm the type of food that was there, so they'd kind of evolved to like suit their surroundings....like over a long time they've kind of changed in like – erm to kind of like make life easier (or something)."
>
> Amy, 15 year-old

> "[Evolution is] like developing things...living things. It's kind of like mutation, it's like when you've got one thing like, and then it's like – gives birth and there's something, I wouldn't say wrong, but different about it, and then somebody else would maybe have something different, and then they'd kind of breed...and you'd have this whole new species this kind of like slight mutation.
>
> [For example] that's the whole thing about the chicken and the egg, isn't it? There's supposed to be something that's nearly a chicken, laid an egg, and it actually was a chicken, so the egg came first, because the egg was laid by something that was nearly a chicken but not quite, and then, the little egg was kind of mutated, so.
>
> [We know this] because there wasn't always humans was there, there was only like kind of little tadpoles swimming about, and they kind of evolved into us...because things aren't always the same...there are like records of things, and then there's not exactly the same things, anymore, they're different...it's really gradual...cause it's like that Charles Darwin person, he went to loads of islands, and he looked at different birds, and some of them had little hard beaks, because all they had to eat was nuts, and some of them had really long beaks, because they had to fish in like really deep flowers to get the nectar out. And so, that's something to do with it."
>
> Jilly, 15 year-old

> "[Evolution is] how animals get the way they are, it's – the strongest, like the survival of the fittest, kind of thing...there's an example I saw in a book, and it's like giraffes may have started off with really short necks, so they would have got all the like horrible leaves at the bottom of the trees, and then a couple would be born with longer necks so they would be able to get the nicer more nutritious leaves at the top, and so...if they mated, then they could have more [offspring] with longer necks, and then the leaves would be, just the shorter ones would die out, 'cause – there wouldn't really be any leaves.
>
> [Human beings evolved from] like, erm, monkeys and primates, but in the theories, everyone started from a single – celled thing...I suppose it's logical, because if you - I don't know, 'cause erm Charles Darwin, he kind of thought of this idea, he studied finches, on some islands, and he noticed that the different finches had different shape beaks on certain islands, and he thought there had to be one finch to start off with, or two finches, and then when they had children, they went off to different islands, and depending on the food, they got different types of beaks...he probably did lots more research as well."
>
> Mandy, 15 year-old

Amy, Jilly and Mandy were all conscientious students, who had elected to take extra science as part of their secondary school course (they were studying 'triple science', whereas the majority of students in the UK only study the standard 'double science' – a broad and balanced science course considered equivalent to two school subjects in the school leaving examinations). All three were quite happy to accept evolution, and all had picked up some general ideas about the topic. They all accepted the variation in finch beaks that Darwin observed at the Galapagos as evidence for evolution, although in itself this is at best suggestive and hardly conclusive. It certainly does not provide a convincing case that all living things evolved from single celled organisms, as Mandy suggests. Jilly accepts the fossil record as evidence, but apparently on trust without knowing any details. She talks about mutations, but without any notion of how selection might act on mutations: her account seems more in line with totally random changes that just happen to lead to present life forms. Mandy reports the giraffe example, apparently misconstruing why reaching higher leaves would even be an advantage, but does not make explicit that for this narrative to 'work' one has to accept both that initially some giraffes would be born with longer necks (without having long-necked parents), and that this characteristic can then be passed on through sexual selection.

I am certainly not looking to criticise these students. They had experienced limited instruction in a topic that was just one of many they had to study on their course, and they all seemed to have acquired something of the gist of the idea. Indeed, sadly, they were likely studying effectively if their aim was to develop sufficient understanding to pass their school examinations. Moreover, like most students in the UK, they had no strong reasons to question this bit of science, and so accepted the arguments with as much (that is, as little) critique as if they had been learning about how the electrical bell works or how ammonia is made industrially.

In some ways it is disappointing that students are so easy persuaded by the authority of science teachers and textbooks, but when evolution is not seen as contentious then this level of understanding is probably a good enough base for developing a more nuanced understanding when the topic is later studied at a higher level (and all three of these girls did opt to study biology or human biology further after reaching the school leaving age). And of course their 'faith' in science teaching was justified: evolution by natural selection is a well supported theory - as Mandy had correctly assumed, Darwin did indeed do "lots more research as well" before asking us to accept natural selection.

But not all students come to science open to accepting ideas about evolution. In the UK context only a small proportion of school children come from 'fundamentalist' religious backgrounds that reject evolution. However, that may not fully reflect the potential magnitude of the issue. A small-scale study with 13-14 years olds from four diverse English schools for the *Learning about Science and Religion* project[2], suggested that some school children see scientific teaching about evolution as contrary to their faith because it seems inconsistent with the stories in Genesis 1 and 2, *even when* they attend mainstream Christian churches (such as Anglican or Roman Catholic churches) that fully accept the scientific accounts of evolution (Taber, 2009b). These pupils seem to just *assume* that they are meant to take the Genesis stories literally, even though that is not actually part of their faith tradition. This is of concern, especially when in some other countries (including the USA), significant proportions of students will come from backgrounds where evolution is actually denied by their faith leaders.

If the level of understanding developed by most secondary school children is similar to that displayed by Amy, Jilly and Mandy (and of course, generalisation from a few cases in one school should not be assumed) then it hardly provides a robust appreciation of the strength of the scientific case for natural selection. Although that scientific case is extremely well supported, without understanding all the links in the argument, nor actually being exposed in some detail to the wide range of evidence, it is clear that school children who have a Worldview inconsistent with evolution are hardly likely to be persuaded to accept the scientific model (Taber, 2009a). This has to be of concern when if there is one scientific theory which, more than any other, needs to be understood to appreciate the importance of astrobiology, it is surely natural selection.

CONCLUSION

The central argument of this chapter has been that if scientists working in astrobiology genuinely want to develop public understanding of their field, then they need to consider moving beyond the notion of 'science communication' and appreciate the challenges faced by professional practitioners of science education. The 'bad news' in this regard is (a) that we know that it can sometimes be very hard to bring about conceptual change when learners have strong preconceptions of a topic, and (b) that when we consider the pre-requisite knowledge for understanding new discoveries in astrobiology, the existing research in science education suggests we are dealing with topics where there are great learning difficulties. Some of the key points highlighted have been that:-

- People generally do not understand the basic mechanics of linear, let alone orbital, motion even after being taught the topic in school;

[2] The *Learning About Science and Religion* project is led by Dr Berry Billingsley of Reading University, through the Faraday Institute for Science and Religion, and is supported by funding awarded by the John Templeton Foundation.

- There are common alternative conceptions about the 'structure' of the universe, with little understanding of the nature of different types of bodies, and limited appreciation of the vastness of scale (in terms of space or time);
- School teaching tends to leave pupils with a dubious understanding of why chemical reactions (a key feature of appreciating the science of the formation and evolution of life) occur;
- Evolution by natural selection is too complex a theory to be readily understood by most school children – although the basic components may be within students' grasp, coordinating them into a coherent model is demanding, and will not be achieved without studying the topic over time, in some depth.

In addition to all this, we also know that within popular culture there are common ideas and perceptions which are likely to distort any new learning in this area: perceptions of 'alien' life are heavily biased towards intelligent humanoid beings; even many people in the USA do not believe that information about the space programme is always true and reliable (i.e. the widespread belief that the moon-landings were a 'hoax'); and many people have a worldview that excludes accepting that life on earth could actually have evolved - a rather major consideration for a science with such a strong focus on how life evolves.

There are clearly major challenges here. However, there are some positive factors that should also be borne in mind. Even if the public tends to see life beyond earth in a rather distorted way, there remains a very strong interest in the issue. The aliens of popular fiction, and even the fascination with UFOs and the associated conspiracy theories, at least provide a strong motivation to learn about space, that is found for few other areas of science outside of medicine. Attempts to build extra-mural courses around astrobiology have been reported to have been successful in both informing members of the public, and enthusing them into advocates for science (Brake, Griffiths, Hook, & Harris, 2006). This should at least be reassuring for scientists attempting to compete with the extraordinary claims of much pseudoscience. For as Carl Sagan (1995) warned, "If you are awash in lost continents and channeling and UFOs and all the long litany of claims so well exposed in the Skeptical Inquirer, you may not have intellectual room for the findings of science. You're sated with wonder."

It has been suggested that the inter-disciplinary nature of astrobiology - described by Cockell (2002: 265) as a "retreat into an inter-disciplinary milieu in the face of questions that perplex us" - makes it an excellent context for helping people understand the nature of science (Oliveira, 2008; Rodrigues & Carrapiço, 2005), and for "the development of more holistic learning and contributing to the flexibility of the students' mental structure" (Rodrigues & Carrapiço, 2006: R-2). Indeed it has been argued that,

> "Astrobiology provides an ideal forum in which to discuss questions relating to the public and cultural placement of science. Astrobiology is a rewarding study of the boundaries between science, art, religion, philosophy and pseudoscience, and also acts as a measure of the dialectic between our advancing human knowledge and that which remains undiscovered."
>
> Brake et al., 2006: 320

The conclusions to be drawn then from this discussion of the public understanding of astrobiology then would be:

1. There are good reasons for scientists to engage with the public and seek to inform people about their professional work;
2. However, the notion of science 'communication' may underplay the very real difficulties of bringing about conceptual change in science, as revealed in a great deal of educational research;
3. Given what we know about the learning difficulties in science topics that astrobiology draws upon, the field of astrobiology presents a particular challenge for effective science communication;
4. Widespread perceptions of space and aliens from popular culture will likely distort the way much of the public understands information about astrobiology;
5. However, the strong interest in space travel and extra-terrestrial life are useful motivators that lead to much interest in the topic; and
6. Effective public education about astrobiology can potentially contribute to wider aims for the public understanding of science, and in particular better appreciating the nature of science.

Taking these points together, the message to scientists working in areas of astrobiology is to indeed prioritise explaining their work to the public, but to be aware of the likely difficulty of effectively communicating the science - and so to prepare themselves for this by drawing upon the literature from science education about learning difficulties and common alternative conceptions in the pre-requisite topics, and from the on-going programme of research into how to best effect conceptual change in science (Taber, 2009a).

REFERENCES

Barbour, I. (2002). *Nature, Human Nature, and God.* London: Society for Promoting Christian Knowledge.

Brake, M., Griffiths, M., Hook, N. & Harris, S. (2006). Alien worlds: astrobiology and public outreach. *International Journal of Astrobiology, 5*(4), 319–324.

Bruner, J. S. (1960). *The Process of Education.* New York: Vintage Books.

Chi, M. T. H. (1992). Conceptual change within and across ontological categories: examples from learning and discovery in science, In R. N. Giere (Eds.) *Cognitive Models in Science,* 129-186. Minneapolis: University of Minnesota Press.

Claxton, G. (2005). *The Wayward Mind: an intimate history of the unconscious.* London: Little Brown.

Cockell, C. (2002). Astrobiology—a new opportunity for interdisciplinary thinking. *Space Policy, 18,* 263–266.

Darwin, C. (1859/1968). *The Origin of Species by Means of Natural Selection, or the preservation of favoured races in the struggle for life.* Harmondsworth, Middlesex: Penguin.

Darwin, C. & Wallace, A. (1858). On the Tendency of Species to form Varieties; and on the Perpetuation of Varieties and Species by Natural Means of Selection. *Proceedings of the Linnean Society, 3,* 45-62.

Dawkins, R. (1988). *The Blind Watchmaker.* Harmondsworth, Middlesex: Penguin Books.

diSessa, A. A. (1993). Towards an epistemology of physics. *Cognition and Instruction, 10*(2&3), 105-225.

Duit, R. (2007). *Bibliography - Students' and Teachers' Conceptions and Science Education*, from http://www.ipn.uni-kiel.de/aktuell/stcse/stcse.html

Eldredge, N. (1995). *Reinventing Darwin: The great evolutionary debate*, London: Weidenfeld and Nicolson.

Engeström, Y., Miettinen, R. & Punamäki, R. L. (1999). *Perspectives on Activity Theory.*, Cambridge: Cambridge University Press.

Fensham, P. J. (2004). *Defining an Identity: The evolution of science education as a field of research*. Dordrecht: Kluwer Academic Publishers.

Frolov, I. T. (1991). *Philosophy and History of Genetics: the inquiry and the debates*. London: McDonald & Co.

Gagné, R. M. & Briggs, L. J. (1974). *Principles of Instructional Design*. New York: Holt, Rinehart & Winston.

Gauld, C. (1989). A study of pupils' responses to empirical evidence. In R. Millar (Ed.), *Doing Science: images of science in science education,* (62-82). London: The Falmer Press.

Gilbert, J. K., Watts, D. M. & Osborne, R. J. (1985). Eliciting student views using an interview-about-instances technique. In L. H. T. West & A. L. Pines (Eds.) *Cognitive Structure and Conceptual Change,* (11-27). London: Academic Press.

Gilbert, J. K. & Zylbersztajn, A. (1985). A conceptual framework for science education: The case study of force and movement. *European Journal of Science Education, 7*(2), 107-120.

Griffiths, M. (2004). Broad horizons – SETI, SF and education. *International Journal of Astrobiology, 3*(2), 175–181.

Halpern, D. F. (2000). Thinking critically about critical thinking: lessons from cognitive psychology. In S. L. Riedel, R. A. Morath & P. T. McGonigle (Eds.) *Training Critical Thinking Skills for Battle Command: ARI Workshop Proceedings,* (22-32). Alexandria, VA: US Army Research Institute for the Behavioral and Social Sciences.

Hammer, D. (2000). Student resources for learning introductory physics. *American Journal of Physics, 68*(7-Physics Education Research Supplement), S52-S59.

Irwin, H. J. (1993). Belief in the paranormal: a review of the empirical literature. *The Journal of the American Society for Psychical Research, 87*(1), 1-39.

Jenkins, E. W. (1979). *From Armstrong to Nuffield: Studies in twentieth-century science education in England and Wales*. London: John Murray.

Jenkins, E. W. (Eds.) (1997). *Innovations in Science and Technology Education, (Vol. VI)*. Paris: UNESCO Publishing.

Jones, R. A. (2004). They came in peace for all mankind: popular culture as a reflection of public attitudes to space. *Space Policy, 20*, 45-48.

Lakoff, G. & Johnson, M. (1980). Conceptual metaphor in everyday language. *The Journal of Philosophy, 77*(8), 453-486.

Luria, A. R. (1976). *Cognitive Development: Its cultural and social foundations*. Cambridge, Massachusetts: Harvard University Press.

Luria, A. R. (1987). *The Mind of a Mnemonist*. Cambridge, Massachusetts: Harvard University Press.

McCloskey, M., Carmazza, A. & Green, B. (1980). Curvilinear motion in the absence of external forces: naïve beliefs about the motion of objects. *Science, 210*, 1139-1141.

Millar, R. & Osborne, J. (1998). *Beyond 2000: Science education for the future*. London: King's College.

Miller, G. A. (1968). The magical number seven, plus or minus two: some limits on our capacity for processing information. In *The Psychology of Communication: Seven essays,* (21-50). Harmondsworth: Penguin.

Nussbaum, J. (1985). The Earth as a cosmic body. In R. Driver, E. Guesne & A. Tiberghien (Eds.) *Children's Ideas in Science,* (170-192). Milton Keynes: Open University Press.

Oliveira, C. (2008). Astrobiology for the 21st Century. *CAPjournal,* (2), 24-25.

Piaget, J. (1929/1973). *The Child's Conception of The World,* (J. Tomlinson & A. Tomlinson, Trans.). St. Albans: Granada.

Piaget, J. (1972). *The Principles of Genetic Epistemology*. London: Routledge & Kegan Paul.

Piaget, J. & Garcia, R. (1989). *Psychogenesis and the History of Science,* (H. Feider, Trans.). New York: Columbia University Press.

Riga, F. (2008). *Students' ideas in astronomy: a study using grounded theory.* Paper presented at the Association for Science Education Annual Conference, Liverpool.

Rodrigues, T. & Carrapiço, F. (2005). *Teaching Astrobiology. A scientific and cultural imperative.* Paper presented at the SPIE conference on Astrobiology and Planetary Missions.

Rodrigues, T. & Carrapico, F. (2006). *How can we teach astrobiology and survive?* Paper presented at the SPIE conference on Instruments, Methods, and Missions for Astrobiology.

Sagan, C. (1995). Wonder and skepticism. *Skeptical Enquirer, 19*(1).

Shayer, M. & Adey, P. (1981). *Towards a Science of Science Teaching: Cognitive development and curriculum demand.* Oxford: Heinemann Educational Books.

Snow, C. P. (1959/1998). The Rede Lecture, 1959: The two cultures. In *The Two Cultures,* (1-51). Cambridge: Cambridge University Press.

Strike, K. A. & Posner, G. J. (1985). A conceptual change view of learning and understanding,. In L. H. T. West & A. L. Pines (Eds.) *Cognitive Structure and Conceptual Change,* (211-231). London: Academic Press.

Taber, K. S. (1998). An alternative conceptual framework from chemistry education. *International Journal of Science Education, 20*(5), 597-608.

Taber, K. S. (2000). Multiple frameworks?: Evidence of manifold conceptions in individual cognitive structure. *International Journal of Science Education, 22*(4), 399-417.

Taber, K. S. (2001). The mismatch between assumed prior knowledge and the learner's conceptions: a typology of learning impediments. *Educational Studies, 27*(2), 159-171.

Taber, K. S. (2002). *Chemical Misconceptions - prevention, diagnosis and cure: Theoretical background, (Vol. 1).* London: Royal Society of Chemistry.

Taber, K. S. (2005). Learning quanta: barriers to stimulating transitions in student understanding of orbital ideas. *Science Education, 89*(1), 94-116.

Taber, K. S. (2006). Teaching about ideas and evidence in science – towards a genuinely broad and balanced 'science for all'. *School Science Review, 87*(321), 26-28.

Taber, K. S. (2008a). Exploring conceptual integration in student thinking: evidence from a case study. *International Journal of Science Education, 30*(14), 1915-1943.

Taber, K. S. (2008b). Towards a curricular model of the nature of science. *Science & Education, 17*(2-3), 179-218.

Taber, K. S. (2009a). *Progressing Science Education: Constructing the scientific research programme into the contingent nature of learning science*. Dordrecht: Springer.

Taber, K. S. (2009b). *Science, God or both?: Secondary learners' reasoning about science and religion*. Paper presented at the CamERA Educational Research Conference, University of Cambridge Faculty of Education, March 2009.

Thagard, P. (1992). *Conceptual Revolutions*. Oxford: Princeton University Press.

The Royal Society. (2005). *Factors affecting Science Communication: Report on qualitative research*.

Toulmin, S. & Goodfield, J. (1962/1999). *The Fabric of the Heavens: The development of astronomy and dynamics*. Chicago: University of Chicago Press.

Vosniadou, S. (Eds.) (2008). *International Handbook of Research on Conceptual Change*. London: Routledge.

Vygotsky, L. S. (1978). *Mind in Society: The development of higher psychological processes*. Cambridge, Massachusetts: Harvard University Press.

Watts, D. M. & Zylbersztajn, A. (1981). A survey of some children's ideas about force. *Physics Education, 16*(6), 360-365.

White, R. T. (1998). Research, theories of learning, principles of teaching and classroom practice: Examples and issues. *Studies in Science Education, 31*, 55 - 70.

In: Astrobiology: Physical Origin, Biological Evolution... ISBN: 978-1-60741-290-8
Editors: S. Hegedus, J. Csonka, pp. 27-57

Chapter 2

WATER INTERFACIAL PROCESSES IN PREBIOTIC CHEMISTRY

Marta Ruiz-Bermejo**, *César Menor-Salván*, *María-Paz Zorzano*, *Zoubir El-Hachemi*, *Susana Osuna-Esteban* and *Sabino Veintemillas-Verdaguer

Centro de Astrobiología [Consejo Superior de Investigaciones Científicas-Instituto Nacional de Técnica Aeroespacial (CSIC-INTA)]. Carretera Torrejón-Ajalvir, Km. 4,2. E-28850 Torrejón de Ardoz, Madrid, Spain.

ABSTRACT

Prebiotic chemistry studies the possible origins of the building blocks of living organisms, as well as their chirality. The hypothesis commonly accepted to explain the origin of life is that simple inorganic molecules, for instance, CH_4, CO_2, N_2, NH_3, reacted together in order to form simple organic compounds that then underwent subsequent reactions to yield biologically functional molecules. These compounds self-organized and self-assembled to increase the organization at molecular levels. This succession of chemical processes is considered the most likely pathway for the emergence of the first living organism. Therefore, the first step to verify this theory is to check how the components of proteins and nucleic acids, among others, could be formed under abiotic conditions. Additionally, the chirality of the constituents of these polymers is another important point. A chiral molecule is a molecule that is not superimposable to its mirror image, L and D configuration. All amino acids in nature have L configuration and all sugars are D. To date, the origin of homochirality in nature is still not understood.

The presence of water is considered a fundamental factor for the development of life. Approximately 72% of the surface of the Earth is covered by liquid water in the form of oceans. This huge water pool and the atmosphere are connected by a cycle of evaporation, condensation, and rain. In addition, oceanic organic and inorganic surface products can be transmitted within aerosols to the atmosphere above, where they are then exposed to radiation and electrical discharges in the form of lightning. The terrestrial

* Corresponding author: E-mail: ruizbm@inta.esPhone: 34 91 520 6402/6458, Fax: 34 91 520 6410.

water cycle and its changes of state and distribution must have played a significant role in the origin of organic molecules under abiotic conditions, as well as in the symmetry breaking.

In this review, we explain some aspects related to aqueous interfacial processes in the context of prebiotic chemistry under the point of view of our current experimental results. We propose a possible role for the cycle of natural aerosols (i.e., from the ocean water pool to the atmosphere and back to the ocean, bubble-aerosol-droplet cycle) and demonstrate the importance of melting-freeze water cycles. These processes may play a relevant role in the synthesis of organic molecules directly related to the first self-organized systems. We also explain a plausible role for the nucleation processes in the solid-vapor interface in the symmetry breaking.

1. INTRODUCTION

The emergence of life is one of the most intricate scientific problems. Prebiotic chemistry studies the possible origins of the building blocks of living organisms starting from abiotic reactions of simple molecules. Thus, prebiotic chemistry tries to explain the abiotic formation of proteins, nucleic acids, sugars, carboxylic acids and lipids as well as the production of analogous or alternative organic compounds, such as polycyclic aromatic hydrocarbons (PAHs) and peptide nucleic acids (PNAs), which were able to constitute the first living matter. In this context, a relevant point is to find the origin of chirality, i.e., the homochirality of natural amino acids (L) and sugars (D) that remains a puzzle for the theories concerning the chemical origin of life. Mirror-image asymmetric molecules, i.e., chiral isomers or enantiomers, are usually considered as chemically identical. In spite of their chemical equivalence, only one enantiomeric form is found in living organisms. On the other hand, water is considered a fundamental factor in the emergence of life. Thus, the terrestrial water cycle and its changes of state and distribution must have played a significant role in the origin of organic molecules under abiotic conditions and in their chiral symmetry breaking.

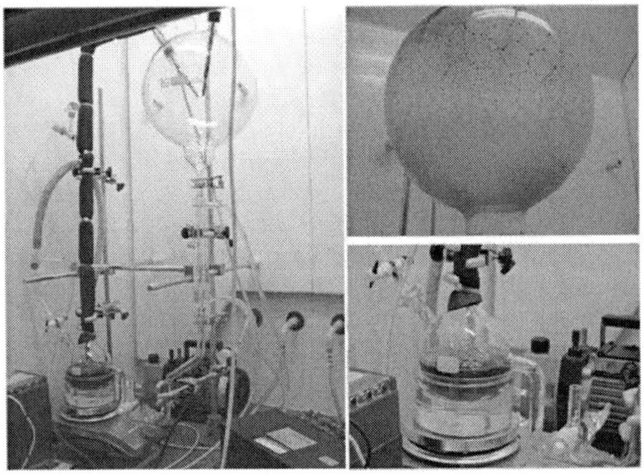

Figure 1. Similar system like the one used in 1953 by S. L. Miller to simulated the conditions of the early Earth and the possible formation of the first biomolecules. The red-brownish solution is formed after one week of spark discharges and several amino acids were detected in it.

There are two fundamental approaches to the study of the origin of life: the top-down approach and the bottom-up approach (Jortner 2006). The first one considers the origin of the constituents of modern biochemistry and their present organization. The second one focuses on compounds that may have been produced under prebiotic planetary conditions and on how they may have been assembled. The key of the study of the origin of life is the overlap between these two approaches. In the line of the bottom-up approach, the Chemical Evolution Hypothesis states that every chemical reactions led to increased complexity and organization of substances resulting in the generation of life. The first step in chemical evolution is the abiotic and spontaneous generation of main bioorganic molecules, such as amino acids, nucleic acid bases, sugars and carboxylic acids, from very simple carbon sources (CH_4, CO_2 or CO). To start these reactions, the atoms that will form these biomolecules (C, H, N, O) have to be in a stable form, for instance, in a gaseous molecular state. An external energy source is required in order to activate these gases and allow them to recombine into a mixture of more complex organic molecules.

Löeb (1913), Growth and Suess (1938) and Garrison et al. (1951) performed some pioneering experiments about the abiotic generation of organic substances from a simple carbon source and an external energy source. In these experiments, simple organic compounds, such as glycine, formaldehyde and formic acid, were identified together with other organic substances of unknown nature. In 1953, Miller proposed a new set-up that led to a breakthrough for experimental studies concerning the abiotic synthesis of bioorganics. Miller simulated a plausible atmosphere of the early Earth ($CH_4 + NH_3 + H_2$) according to the theory of Urey (1952) and demonstrated that electric discharges supplied to a strongly reducing mixture could successfully generate amino acids in a relatively high yield (Miller 1953). A system similar to the one used by Miller is shown in Figure 1. A distinctive feature of this experimental set-up is the continuous recycling of water in the gas phase in the reaction chamber.

Using this system and after one week of spark discharges on the reducing atmosphere, Miller detected the following amino acids in the red-brownish solution formed: glycine, alanine, aspartic acid, β-alanine and α-aminobutyric acid. Glycine, alanine and aspartic acid are present in proteins. This was the starting point of research studies in prebiotic chemistry.

Following this experiment, Miller and others studied different gas mixtures and different sources of energy. Several experiments focused on a different initial gas mixture to emulate the primitive Earth atmosphere, the atmospheres of the other planets, as well as the composition of the interstellar medium have been made. In the environment of the early Earth, or in ancient or other actual planetary atmospheres, the plausible energy sources are: solar radiations, electric discharges (it is believed that the storms were continuous and numerous in the early Earth), volcanic heat, radioactivity, cosmic rays, and meteorite impacts (Kobayashi et al. 2001). For the interstellar medium, common energy sources are ultraviolet radiations and cosmic rays (Muñoz-Caro et al. 2001, Mennela et al. 2001). Finally, the pressure and temperature conditions are also different for other planets of the solar system, different terrestrial environments, and the interstellar medium. Thus, prebiotic conditions that can be simulated in the laboratory are numerous and the production of organic compounds of biological interest generally depends directly on the prebiotic experimental conditions. However, in spite of the numerous experiments carried out, the diversity of the bioorganic compounds obtained is very limited. In most prebiotic experiments, amino acids are formed but their diversity and yield depend on the gas mixtures (ratios C/H and C/N) and the energy source (see, e.g., Schlesinger and Miller 1983). It is noteworthy that all mixtures of amino acids obtained in these experiments are racemic and that the formation of amino acids containing sulfur and aromatic amino acids remains unclear. Nucleic acid bases and short carboxylic acids (no more than six carbons) have only been detected in a few experiments. No sugars have been clearly identified in experiments from simple carbon sources and external energy sources.

Prebiotic experiments as described above with gas mixtures and vapor water (see e.g., Utsumi et al. 2002; Miller 1955) or liquid water (see e.g., Takahashi et al. 2005; Plankensteiner et al. 2004; Takano et al. 2003) as external energy source have been reported but only one example with an extremely low amount of water (less than 10 ppm) has been published (Miyakawa et al. 1998). In others examples, such as in experiments of irradiation of interstellar ice analogues (see e.g., Bernstein et al. 2002; Muñoz-Caro et al. 2002; Kobayashi et al. 1995) or in an experiment using cold plasma at a low temperature (see e.g., Simionescu et al. 1974), solid water was used. However, approximately 72% of the surface of the Earth is covered by liquid water in the form of oceans. This huge water pool and the atmosphere are connected through a cycle of evaporation, condensation, and rain. The presence of water in other planetary environments has also been suggested (Kargel 2006, Kivelson et al. 2000). It is now particularly clear that ice water exists on Mars on the polar caps (Bibring et al. 2004, Kieffer and Titus 2001) and its regolith (Zent et al. 2009, Boynton et al. 2002, Feldman et al. 2002, Fanale et al. 1982). Traces of water vapors were also detected in its atmosphere (Fouchet et al. 2007, Encrenaz et al. 2005, Smith 2002).

Taking into account the peculiar characteristic of water, it has been speculated that water in the form of aerosol could have played an important role as prebiotic microreactors in the formation of the first bioorganic compounds and in the origin of homochirality (Donaldson et al. 2004, Tuck 2002, Dobson et al. 2000). Indeed, there is a continuous recycling process that supports the global mixing of products in water aerosols. Additionally, there is an increased proportion of interfaces with the liquid present in the atmosphere above the ocean due to the continuous aerosol bursting processes. The aerosol bursting process forms droplets that are rich in liquid bulk surface products. At these aqueous interfaces, evaporation takes place rapidly. As the aerosols evaporate, the solutes on their surface concentrate, the air-water interface enhances the effective molecular ordered packing and increases the local concentration and the influence of short-range forces (Zepik 2002). In particular, hydrophobic molecules will arrange in such a way that their hydrophobic side sticks out of the liquid, and other inorganic products, such as salts, may be also be present in the form of ordered crystals. Moreover, there is a variety of physical-chemical micro-environments in such a case. For instance, liquid surfaces are very rich in hydroxyl ions, the liquid-vapor phase transition processes at the water-air interfaces may induce molecular reactions associated with dehydration, the solutes dissolved in water change in nature (salts, minerals, metals) and concentration depending on the micro-environment, aerosols and atmospheric products are exposed to the ionizing radiation and electrical discharges, and finally at a planetary scale, there are a strong thermal and pressure gradients from the bottom of the ocean to the surface of the liquid, as well as from the bottom to the top of the atmosphere. Thus, the terrestrial water cycle provides a wide range of possible experimental conditions and physico-chemical environments to explore prebiotic chemistry and symmetry breaking.

The possible role of water aerosols as important reactors in prebiotic chemistry was first proposed by Shah (1970). This suggestion is supported by physical-chemical observations that demonstrate that the formation of an aerosol only depends on a liquid water-air interface and a physical mechanism that ejects bubbles into the atmosphere, such as wind, sea waves or shock waves (Ellison et al. 1999). On modern Earth, aerosols are ubiquitous in the troposphere and play an important role in climate and atmospheric chemistry (Ellison et al. 1999). Laboratory work (Donaldson and Anderson 1999) has shown that small organic molecules tend to concentrate at the water-air interface. Similarly, observations of the composition of natural aerosol particles have shown that at least 10% of the aerosol is organic matter, with more than 10.000 peaks in gas chromatography (Murphy 2005 and Murphy et al. 1998) and organic coating of surfactants in marine aerosol particles (Tervahattu et al. 2002). Due to these facts, we could assume that the bubble-aerosol-droplet cycle (the bursting of bubbles that injects into the atmosphere the aerosol particles and subsequent condensation in droplets) was active in the Archean epoch (Figure 2). This process could have been functionally relevant to the chemical evolution cycle on the early Earth or elsewhere. It is important to note that aerosol particles are different from cloud droplets and raindrops. The latter are larger hydrometeors with droplet radii greater by 1 to 4 orders of magnitude than aerosol droplets. They also have a small fractional organic content compared to aerosol particles (Tuck 2002).

Finally, the study of prebiotic chemistry in cold environments has recently gained significance with the investigation of the icy moons of Saturn. The possible existence of liquid water plumes (with eruption of methane, carbon dioxide, and nitrogen), recycled ice in Enceladus, and a liquid briny ocean in Europa suggest the presence of a water freeze-melt cycle (Kargel 2006, Kivelson et al. 2000). The Cassini observation mission suggested the presence of a subsurface liquid ocean of H_2O/NH_3 in Titan (Lorenz et al. 2008). In the case of the early Earth, it could have been cold enough to freeze at least in certain regions. On the other hand, it is well known that when aqueous solutions are frozen, i.e., when the ice lattice forms, solutes are excluded and extremely concentrated brines may be formed. Additionally, the freeze cycling process is a unique environment that generates a concentrated liquid phase with strong electric and temperature gradients and high pressure providing an efficient reactor (Trinks et al. 2005). These facts motivate experiments of prebiotic chemistry simulating water freeze-melt cycles.

2. Pure Water Aerosols: Formation of Organic Substances

Differences in reactivity between the bulk water in any physical state and the water present in air-water interfaces are expected because of increasing superficial phenomena. This fact is due to the effective concentration of organic reagents that can be greatly increased at the surface and to the fact that the reaction rates are accelerated and the products are modified by reaction with their solution counterparts. Reactions that are not favorable in the bulk solution may occur at atmospheric aqueous interfaces (aerosols, lakes, oceans) and these may have contributed to the formation of biopolymers in a non-enzymatic prebiotic scenario (Donaldson and Vaida 2006).

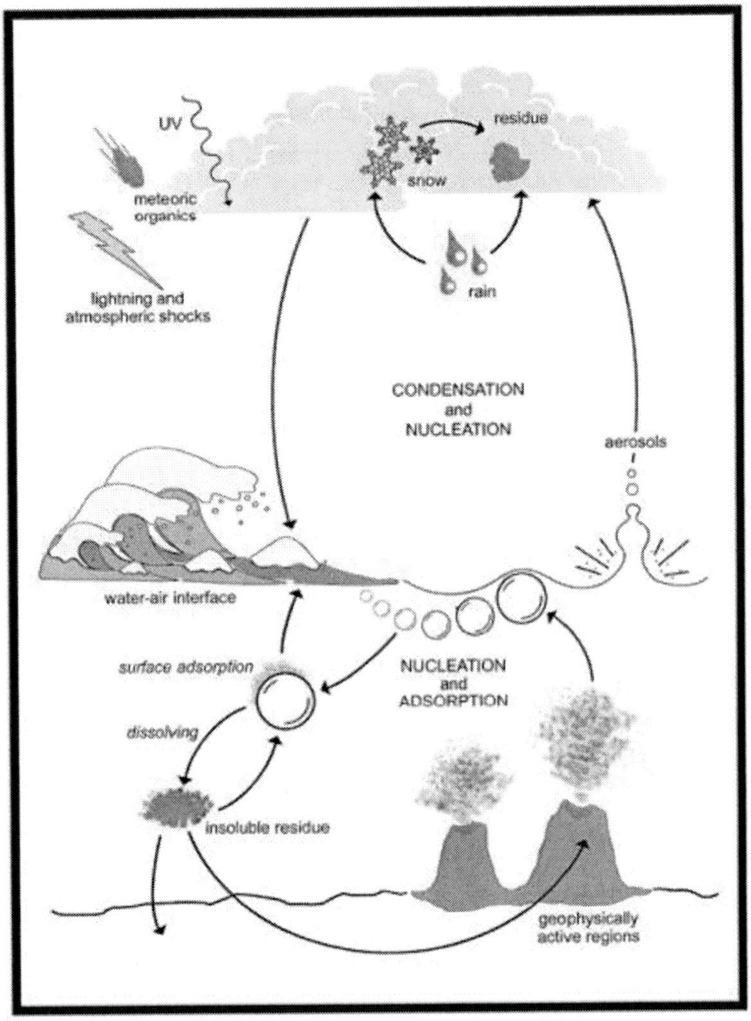

Figure 2. Bubble-aerosol-droplet cycle: Bursting of bubbles that injects into the atmosphere the aerosol particles and subsequent condensation in droplets. The bubble-aerosol-droplet cycle could be active in the Archean epoch. (Lerman, L. Lunar and Planetary Science 2006).

In order to test this hypothesis and to check the formation of bioorganic compounds, a bubble-aerosol-droplet cycle was experimentally generated in plausible prebiotic conditions (see figure 3). As a control, an experiment was performed with liquid water and under the same conditions but without aerosols. The aerosol was generated by low power ultrasonic transductors. These devices are extensively used in the simulations of aqueous aerosols (King et al. 2004) and in pharmaceutical sciences (Steckel and Eskandar 2003). Although the mechanism of aerosols should be effective in atmospheric models of a large range of compositions, a mixture of CH_4, N_2 and H_2 was chosen because this composition is known to yield amino acids in amounts sufficient for comparative quantitative analysis (Schlesinger and Miller 1983).

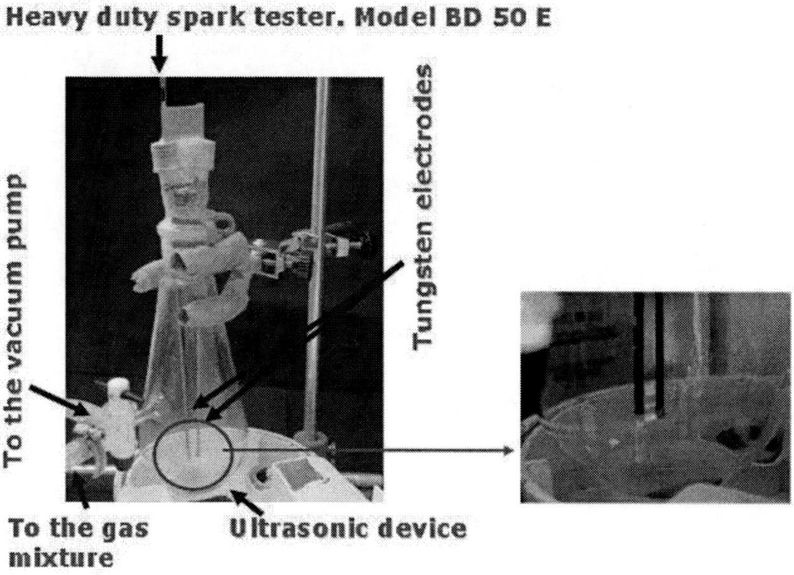

Figure 3. System to simulated the bubble-aerosol-droplet cycle. The reactor is set up in an ultrasonic device that generates the aerosols, by focusing the ultrasound wave (1.8 MHz, 30 W) onto the liquid surface.

Under these conditions, the spark discharge leads to the formation of organic material in different non volatile phases: a soluble fraction that remains in the liquid water pool as a yellow-brown solution of pH 8.6 (hydrophilic *tholin* after the solution has been freeze-dried) and an insoluble fraction (hydrophobic *tholin*) isolated by centrifugation. The term *tholin* is used to name the complex organic material that is obtained by irradiation of atmospheres that contain CO, CO_2 or CH_4, as carbon sources, and N_2 and NH_3, as nitrogen sources (Sagan et al. 1993). The control experiment carried out without aerosols yields a soluble fraction (pH = 8.7, hydrophilic *tholin*) and an insoluble fraction (hydrophobic *tholin*) that is colleted as a thin film from the surface of the liquid and from the wall of the reaction vessel. The amount of organic matter obtained in each experiment and the elemental analysis for each *tholin* are shown in table 1.

These data indicate that: i) Aqueous aerosols improve the formation of an organic soluble material. Indeed, the amount of insoluble material collected is small compared to the control experiment. Therefore, the presence of aqueous aerosols leads to a greater amount of water-soluble organic materials than the control experiment with liquid water. ii) The empirical formulas for the hydrophilic *tholins* are similar but they are really different respect on the hydrophobic *tholins*. These differences could be related to the presence of a higher number of functional groups in the hydrophilic *tholins* and to the higher grade of unsaturation of hydrophobic *tholins*.

Chemical analysis of the hydrophilic *tholins* obtained during the experiments with and without aerosols was performed in order to check the effect of the aerosols on the synthesis of organic molecules of biological interest (Ruiz-Bermejo et al. 2007). The compounds identified in each experiment are shown in table 2.

Table 1. C and N fixed and other data of the organic material obtained in aqueous aerosols experiment and control experiment with liquid water. Values are expressed as mean ± SEM of ten experiments perfomed in aerosol and control conditions.

	Hydrophilic *tholin* (Aerosols experiment)	Hydrophobic *tholin* (Aerosols experiment)	Hydrophilic *tholin* (Control experiment)	Hydrophobic *tholin* (Control experiment)
pH (solution)	8.65	Insoluble	8.73	Insoluble
Amount obtained (mg)	60.7 ± 4.4	17.8 ± 1.9	43.3 ± 3.1	25.8 ± 2.8
C fixed (% control)	16.34 ± 1.35	10.48 ± 0.34	16.94 ± 1.21	18.46 ± 2.07
N fixed (% control)	5.98 ± 0.52	0.25 ± 0.03	10.93 ± 0.78	0.99 ± 0.11
C/N	1.8	43	1.8	21.2
C/O	1.8	5.4	1.8	7.6
Empirical formula	$C_9H_{18}N_5O_5$	$C_{30}H_{45}NO_3$	$C_9H_{16}N_5O_5$	$C_{106}H_{157}N_5O_{14}$

Table 2. Identification and quantification of organic molecules with biological interest identified in spark discharge experiments with or without water aerorols. *d* detected (Noise/Signal > 0.1).

		Aqueous aerosols experiment	Liquid water experiment
		Amount (μmol)	Amount (μmol)
AMINO ACIDS			
Glycine	$C_2H_5NO_2$	3.93	2.89
Aminomalonic acid	$C_3H_5NO_4$	1.06	0.001
Alanine	$C_3H_7NO_2$	12.98	8.68
β-Alanine	$C_3H_7NO_2$	1.16	0.21
Sarcosine	$C_3H_7NO_2$	0.45	1.01
Isoserine	$C_3H_7NO_3$	0.34	0.12
Serine	$C_3H_7NO_3$	0.14	0.07
Aspartic acid	$C_4H_7NO_4$	0.34	0.10
Iminodiacetic acid	$C_4H_7NO_4$	0.03	d
2-Aminobutyric acid	$C_4H_9NO_2$	0.18	0.40
3-Aminobutyric acid	$C_4H_9NO_2$	0.68	1.02
3-Aminoisobutyric acid	$C_4H_9NO_2$	1.65	1.34
N-Methylalanine	$C_4H_9NO_2$	d	0.09
Glutamic acid	$C_5H_9NO_4$	0.08	0.16
Ornithine*	$C_5H_{12}N_2O_2$	0.43	0.24
Histidine*	$C_6H_9N_3O_2$	d	-
t-Leucine	$C_6H_{13}NO_2$	0.19	-
TOTAL[a]		**1.63**[a]	**1.19**[a]

Table 2. (Continued)

		Aqueous aerosols experiment	Liquid water experiment
		Amount (μmol)	Amount (μmol)
HYDROXY ACIDS			
Glycolic acid	$C_2H_4O_3$	0.17	-
2,3-Hydroxypropanoic acid	$C_3H_6O_4$	0.02	d
Malic acid	$C_4H_6O_5$	0.03	d
Tartaric acid	$C_4H_6O_6$	0.01	-
2-Hydroxypentanodioic acid	$C_5H_8O_5$	0.01	-
6-Hydroxycaproic acid	$C_6H_{12}O_3$	0.02	-
TOTAL[a]		**0.018**[a]	**d**
CARBOXYLIC ACIDS			
2-Methylbutanodioic acid	$C_5H_8O_4$	0.02	d
Succinic acid	$C_4H_6O_4$	0.05	0.01
Glutaric acid	$C_5H_8O_4$	d	-
Tricarballylic acid	$C_6H_8O_6$	0.01	-
TOTAL[a]		**0.007**[a]	**$8x10^{-5}$**
HETEROCYCLES			
Parabanic acid	$C_3H_2N_2O_3$	0.02	-
Hydantoin	$C_3H_4N_2O_2$	0.06	-
Adenine	$C_5H_5N_5$	0.11	d
2,6-diaminopurine	$C_5H_6N_6$	0.02	-
TOTAL[a]		**0.022**[a]	**d**

[a] yield (%) respect to the total carbon in the system.

Hydrophilic *tholins* were analyzed for their content in amino acids. A set of amino acids was detected after acidic hydrolysis. The amount and diversity of amino acids were greater in the experiment with aerosols than in the control experiment with liquid water. Thus, water aerosols increase the production of amino acids and their diversity. It is important to indicate that no amino acid was detected before hydrolysis. Thus, amino acids are in a precursor form in the crude reaction that will be discussed later.

Heterocycles were also detected. Analysis of the hydrophilic *tholin* from the experiment with aqueous aerosols indicated the formation of parabanic acid, hydantoin, adenine, and 2,6-diaminopurine. These compounds were not found in the control experiment. It is interesting to note that free adenine was obtained in the solution and not one of its unknown precursor, as stated in previous work (Yuasa et al. 1984). Moreover, adenine was obtained in an amount comparable to polymerization experiments of ammonium cyanide solutions (Miyakawa et al. 2002a, b; Levy et al. 2000). We propose a mechanism for the formation of adenine similar to the one previously proposed by Yuasa et al. (1984), in which HCN is involved. The mechanism of formation of adenine and the relatively high yield in adenine obtained in our experiment with aerosols are due to the local enrichment in HCN and diaminomaleodinitrile, or its immediate precursors, on the liquid-gas interface. Moreover, 2, 6-diaminopurine was detected and the synthesis of both purines by polymerization of HCN solutions is consistent (Borquez et al. 2005). Nevertheless, a radical mechanism for the formation of heterocyles cannot be excluded since the aqueous aerosols seem to improve the stabilization of polar radicals (Ruiz-Bermejo et al. 2007b). Cyanoacetylene was produced in good yield from electric discharges acting on $CH_4 + N_2$ mixtures (Sanchez et al. 1966), and Ferris et al. (1968) obtained heterocycles from cyanoacetylene and urea trough radical pathways.

In addition, we have found dihydroxy acids in the aerosol experiments (absent in the control experiment without aerosols). We report on the prebiotic synthesis of 2,3-dihydroxybutanedioic acid (tartaric acid) and 2,3-dihydroxypropanoic acid (glyceric acid). These compounds have not yet been identified in other experiments with spark discharges. Other biologically relevant molecules identified in the products of spark discharges in an environment containing aerosols are propane-1,2,3-tricarboxylic acid (tricarballylic acid), succinic acid, and malic acid. Succinic acid and malic acid are members of the Krebs cycle and the tricarballylic acid is structurally related to citric acid and isocitric acid that are also members of the Krebs cycle. It has been suggested that these carboxylic acids, including tricarballylic acid, could take part in a primordial variant of the Krebs cycle (Melendez-Hevia et al. 1996). However, the hypothesis on self-organizing protometabolic cycles is a controversial issue (Orgel, 2000). Our experiment indicates that aerosol chemistry could have contributed to the production of the raw materials necessary for this suggested primordial metabolism, similarly to previous work on the synthesis of dicarboxylic acids by irradiation of acetic acid solutions (Negron-Mendoza and Ponnamperuma, 1976).

In summary, aerosols increase the prebiotic yield of amino acids and they form an appropriate environment for the formation of purines and other heterocycles, as well as hydroxy- and carboxylic acids.

The differences between the organic compounds detected in the hydrophilic *tholin* from the experiment with aerosols and the *tholin* from the control experiment can be explained by their structural differences and by the different mechanism implicated in their formation. In order to determine the functional groups present in the hydrophilic *tholins* obtained from the experiment with and without aqueous aerosols, several spectroscopic techniques were used: IR spectroscopy, ^{1}H NMR, ^{13}C NMR, XPS spectroscopy, and UV-vis spectroscopy. The functional groups that were identified in both *tholins* are shown in Table 3 (Ruiz-Bermejo et al. 2008).

Table 3. Functional groups present in the hydrophilic tholins from aqueous aerosol experiment and control experiment. The functional groups were identified by the complementary spectroscopic techniques.

Aerosol Experiment	Control Experiment
$-CH_3$, $-CH_2-$, -CH-	$-CH_3$, $-CH_2-$, -CH-
-C=C- (no aromatic)	-C=C- (no aromatic)
-COOH, -COOR	-COOH, -COOR
-OH	-OH
$-NH_2$, -NH-	$-NH_2$, -NH-
-C≡N	-C≡N
-N= / -N=C (heterocycles)	
-NH-C=O (amides)	

It is interesting to point out that carboxylic acids are present in larger amounts in the hydrophilic *tholin* from aqueous aerosols than in the control experiment, as determined by GC-MS and other spectroscopic results. The higher number of resonances in the ^{13}C NMR spectrum and the continuous signal among 3350 and 1700 cm^{-1} in the IR spectrum confirm this fact. Moreover, the XPS data confirm the presence of N= / -N=C and -NH-C=O groups only in the hydrophilic *tholin* from the aerosols experiment. These functional groups are directly related to the previously mentioned heterocycles. The classical Bradford assay for peptide reactivity was negative and the presence of -NH-C=O in *tholin* from the aerosols experiment was attributed to the heterocycles, hydantoin and parabanic acid, since XPS cannot discriminate between a linear and cyclic structure.

Hydrophylic *tholins* were also analyzed by reverse-phase HPLC in order to separate their possible components according to their polarities. The HPLC chromatograms seemed to indicate that both *tholins* are similar complex substances. The differences in polarities of the different fractions were reflected in the UV-vis spectra (Figure 4). The absorptions in the range of 240-290 nm were attributed to conjugated double bonds or aromatic structures, and the *tholin* from the control experiment seemed to contain a higher number of unsaturated bonds than the *tholin* from the aerosols experiment.

The fractionation of the *tholins* according to molecular weight by ultrafiltration and dialysis, using centrifugal filter devices and cellulose acetate membranes for biodialysis, indicated that the aqueous aerosols favored the formation of lighter materials (Table 4).

The molecular weight distribution and the analysis of the organic molecules indicate that the hydrophilic *tholins* are mainly formed by discrete molecules and small polar oligomers, as well as by polymers/heteropolymers with a remarkable molecular weights. The molecular weight of these polymers/heteropolymers is about 28kDa. These measures were made using electrophoretic techniques (Ruiz-Bermejo et al. 2008). Taking into account the spectroscopic data, we propose that these polymers/heteropolymers were constituted by hydrocarbon chains containing a significant number of unsaturated bonds and chargeable polar groups such as –OH, -NH and –COOH. We also propose that the structure of the heavier fraction is similar to the one reported by McDonald et al. (1991).

The nature and ratio between the discrete molecules and the polymers/heteropolymers detected in the hydrophilic *tholins* suggest different mechanisms of formation that compete with each other. The aqueous aerosols favored the formation of light polar compounds and molecular diversity, but partially inhibited the formation of polymers/heteropolymers. In this regard, we propose a set of three-pot reaction mechanisms: one on the bulk water, one in the water-gas interface, and the last one in the gas phase. The hydrophilic character was clearly enhanced by the increase of the water-gas interface. This suggests a mechanism, in which a hydrocarbon chain is formed in the gas phase by following a radical pathway. This chain is located on the surface that contains polar radicals from the gas phase. This could explain the molecular weight distribution in the *tholin* from the aerosols experiment since, under gas-surface enhanced conditions, a higher density of polar groups is present. In fact, these groups (for example, the carboxylates) are terminal groups and favor oligomers or small polymers. In addition, the fast incorporation of hydrophilic groups favor the solubilization of the molecule/polymer that stops growing. The formation of a considerable amount of carboxylic acids in the experiment with aerosols supports this hypothesis. It is also in agreement with the results reported by other authors (Donaldson et al. 2004, Tuck 2002, Ellison et al. 1999). These authors indicated that carboxylic acids were produced by the reaction of OH with hydrocarbons at the water-gas interface, especially in aerosols. In this context, the presence of aqueous aerosols did not lead to the formation of PAHs since the enhanced interfacial attachment of polar groups actively competed with the formation of unsaturated compounds that could polymerize by a radical pathway (Wilson and Atreya 2003).

Another important aspect of the mechanism to consider is that there are one hydrophilic *tholin* and one hydrophobic *tholin* formed instead of a *tholin* with a mixed nature in our experiments. This result is important for the origin of the first membranes because it is clear that the self- assembly phenomenon responsible for the membrane formation is only possible when both hydrophobic and hydrophilic molecular components are present.

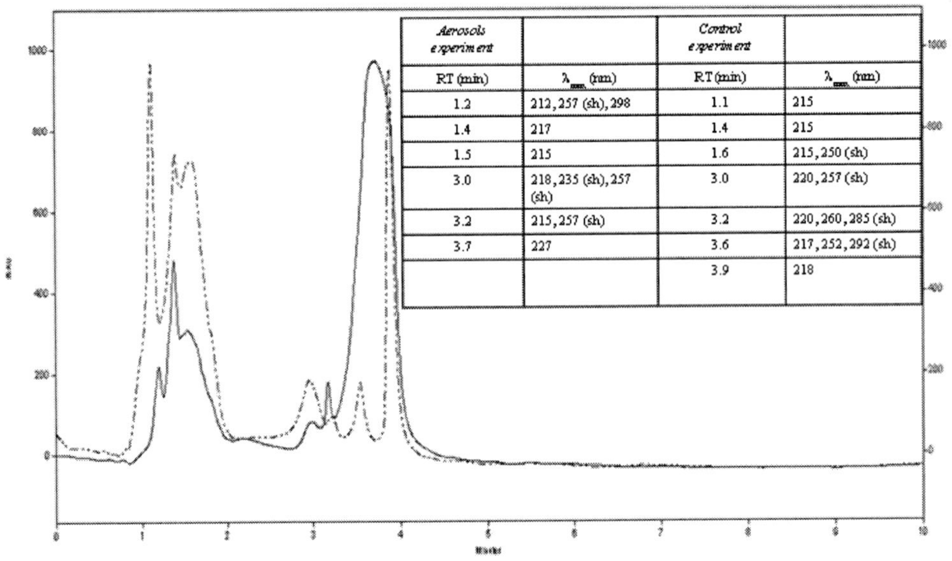

Figure 4. Separation of fractions by reverse-phase HPLC of the hydrophilic *tholins*. The table shows the wavelengths of the absorbance spectra of each fraction.

Table 4. Fractions separated by ultrafiltration and dialysis using filter devices and celluluose acetate membranes.

	Aerosols experiment	Control experiment
Fraction	%	%
< 0.5 kDa	67.8	52.7
0.5 -1 kDa	4.6	2.0
1-3 kDa	16.3	34.8
> 3 kDa (soluble)	7.2	7.5
> 3kDa (non soluble)	4.0	3.0

3. Saline Water Aerosols: Effects on the Formation of Bioorganic Compounds

Pure water has been used in the majority of prebiotic chemistry simulation experiments. Only a few studies describe the effect of saline solutions on this type of experiment (Plankensteiner et al. 2006; Manolache et al. 1997; Kobayashi et al. 1990). Nonetheless, it has been postulated that the ancient sea had a salinity 1.5 to 2 times higher than the current value (Knauth 2005; Morse and Mackenzie 1998) and a pH = 4 – 10, although the pH value of early ocean water is under debate (see e.g., Kempe and Kazmierczak 2002, Morse and Mackenzie 1998, Russell and Hall 1997, Williams and Frausto da Silva 1996, Bada et al. 1994, Macleod et al. 1994, Grotzinger and Kasting 1993, Kasting 1993). The presence of banded iron formations (Poulton et al. 2004, Barley et al. 1997, Holland 1989, Warlker and Brimblecombe 1985) strongly suggests that dissolved iron (ferrous ion) was present in significant quantities in the ocean water under the reducing atmosphere of the Archean epoch. Finally, it has been hypothesized that the concentration of iron could decrease by precipitation as iron carbonate (siderite) (Macleod et al. 1994) or iron sulfide (pyrite) (Summers 2005).

The effects of aerosols and salinity on the prebiotic synthesis of organic compounds are probably correlated. The salinity and pH of the aqueous phase may have an important influence on gas-liquid interfaces where relevant prebiotic reactions are likely to occur (Tervahattu et al. 2004). It has been proposed that a solid core, organic or inorganic, formed by partial evaporation of the saline aerosol droplets, could improve the catalytic properties of aerosol droplets (Lerman and Teng, 2004).

The following solutions were use to produce saline water aerosols in order to test the effects of inorganic salts and ferrous ion on the abiotic formation of organic molecules: i) High salinity solution 1.5 times the concentration of actual sea water adjusted to a pH of 5.8 with HCl (Morse and Mackenzie 1998): 0.63 M NaCl, 0.18 M $MgCl_2$, 0.015 M $CaCl_2$, 0.015 M KCl, 0.02 M $NaHCO_3$. ii) A suspension of FeS in water: 0.8 mg/mL FeS, pH = 7. iii) A suspension of FeS in the same solution as in the first experiment: 0.63 M NaCl, 0.18 M $MgCl_2$, 0.015 M $CaCl_2$, 0.015 M KCl, 0.02 M $NaHCO_3$ + 0.88 mg/mL FeS. iv) A dilute solution of $FeCl_2$ (Mahajan et al. 2003): 0.01 M $FeCl_2$, pH = 7. v) A suspension of siderite in water: 2 mg/mL $FeCO_3$ (natural siderite), pH = 7 (Ruiz-Bermejo et al. 2007b).

Acidic hydrolysis of all non-volatile water-soluble fractions of the experiments leads to a series of amino acids. The amount and number of amino acids detected in the first experiment are greater than in the control experiment with pure water (39.77 μmol compared to 23.64 μmol). The total amount of carbon introduced in both experiments was 126 mg. Consequently, the presence of chloride and bicarbonate salts seems to improve the production of amino acids. A similar result was previously obtained in simulation experiments using proton irradiation, CH_4/N_2 atmosphere and solutions of chlorides and $Fe(NH_4)_2(SO_4)_2$ at pH = 7 (Kobayashi et al. 1990). The presence of insoluble or of low solubility salts, such as FeS and $FeCO_3$, does not affect significantly the synthesis of amino acids (26.01 and 24.19 μmol, respectively). The presence of FeS leads to the formation of amino acids containing sulfur. Cystine was tentatively detected in the second and third experiments. Nevertheless, in the experiment with soluble Fe^{+2}, the total amount of amino acids detected is significantly lower than in the control experiment (10.52 μmol). This is due to the formation of the insoluble Prussian Blue salt [ferric hexacyaneferrate (II), $Fe_4[Fe(CN)_6]_3 \cdot 15\ H_2O$, crystal structure reported by Buser et al.(1977)] from soluble Fe^{+2} and HCN. HCN is one of the main products in prebiotic synthesis experiments using spark discharges and CH_4 (Stribling and Miller 1987) and it is known that when solutions of Fe (II) and CN^- are mixed, $Fe_x(CN)_y$ precipitates are formed (Sharpe 1976). Thus, the precipitation of Prussian Blue consumes carbon and nitrogen from the system. In particular, the formation of amino acids in experiments with spark discharges and CH_4 is thought to follow a mechanism similar to the one of the Strecker synthesis (Miller 1957). In such a process, any reduction in HCN concentration leads to a reduction in the yield of the amino acids. The correlation observed between the decrease of amino acid production and the removal of HCN by precipitation to form Prussian Blue is in agreement with this hypothesis. Our results also suggest that metal-bound cyanide is unable to participate in the Strecker synthesis, apparently contradicting the results of Summer and Lerner (1998). These authors have indicated that the form of cyanide used, i.e., free or bound to Fe(II), has no effect on the Strecker synthesis. In summary, we propose that the mechanism of the Strecker synthesis is the most favorable mechanism for the formation of amino acids under our experimental conditions, but that it is not the only one. Aqueous aerosols favor the formation of low molecular weight compounds, such as aminonitriles, and improve the amino acids yield. Aqueous aerosols also allow the formation of molecules, such as hydantoin, increasing the yield of amino acids and offering an alternative mechanism for the production of amino acids (Taillades et al. 1998).

We have detected hydroxy acids and carboxylic acids in all experiments with saline aerosols. Biologically interesting compounds, such as tartaric acid and glyceric acid, as well as succinic acid, malic acid, and tricarballylic acid were identified in all experiments (see table 5). These molecules were also detected in our control experiment with pure water aerosols, but the presence of dissolved salts further increases the total amount of these products (see Table 5). On the other hand, there is not a remarkable difference in the formation of hydroxy acids and carboxylic acids in the experiment with soluble Fe^{+2}.

The insensitivity of hydroxy and carboxylic acids to Prussian Blue precipitation indicates that HCN may not be implicated in the formation of these compounds. In accordance with previous observations, we propose that the formation of hydroxy acids and carboxylic acids under the conditions assayed follows a radical pathway. This hypothesis is in agreement with the results reported by Donaldson and coworkers (Donaldson et al. 2004). Indeed, they have shown that carboxylic acids are produced by reaction of OH with hydrocarbons at the air-water interface, especially in aerosols. On the other hand, Tuck (Tuck 2002) has argued that chain lengthening in hydrocarbons in the anaerobic prebiotic atmosphere may have resulted from a gas phase reaction with methane, and with further oxidation to alcohols, aldehydes, and acids. Thus, the presence of aqueous aerosols and salts may stabilize radical intermediates that are then more likely to evolve and form the compounds described above.

Table 5. Analysis results of hydroxy acids and carboxylic acids from experiments of saline water aerosols and spark discharges. Experiment 1) Saline solution 1.5 times the concentration of actual sea water, pH = 5.8. Experiment 2) Suspension of FeS. Experiment 3) Suspension of FeS in the same solution as experiment 1. Experiment 4) Solution of $FeCl_2$. Experiment 5) Suspension of siderite.

		Experiment 1	Experiment 2	Experiment 3	Experiment 4	Experiment 5	Water aerosol
HYDROXY ACIDS (μmol)							
Glycolic acid	$C_2H_4O_3$	0.507	0.011	0.478	0.327	0.140	0.17
2,3-Dihydroxypropanoic acid	$C_3H_6O_4$	0.017	0.002	0.010	0.008	0.028	0.02
Malic acid	$C_4H_6O_5$	0.018	0.029	0.007	0.011	0.019	0.03
Tartatic acid	$C_4H_6O_6$	0.019	0.003	0.003	0.003	0.044	0.01
2,3-Dihydroxybutanoic acid	$C_4H_8O_4$	0.003	-	0.004	0.007	-	-
2-Hydroxypentanodioic acid	$C_5H_8O_5$	-	-	0.002	-	d	0.01
2,4-Dihydroxypentanodioic acid	$C_5H_8O_6$	-	-	-	-	d	
Hydroxycaproic acid	$C_6H_{12}O_3$	0.015	0.001	0.015	0.003	0.055	0.02
2,3-Dimethyl-3-hydroxyglutaric acid	$C_7H_{12}O_5$	-	0.007	-	-	0.009	
Total		**0.58**	**0.05**	**0.52**	**0.36**	**0.30**	**0.26**
CARBOXILIC ACIDS (μmol)							
Succinic acid	$C_4H_6O_4$	0.040	0.020	0.032	0.028	0.187	0.05
But-2-enedioic acid	$C_4H_8O_4$	-	0.003	0.011	-	0.013	0.02
2-Methyl maleic acid	$C_5H_6O_4$	-	-	-	-	d	-
2-Methylsuccinic acid	$C_5H_8O_4$	0.003	0.002	0.003	0.002	0.020	-
Tricarballylic acid	$C_6H_8O_6$	0.003	0.002	0.001	0.003	0.030	0.01
2-Methylglutaric acid	$C_6H_{10}O_4$	0.002	-	-	-	d	-
Adipic acid	$C_6H_{10}O_4$	0.001	-	-	-	-	-
Total		**0.05**	**0.03**	**0.05**	**0.03**	**0.08**	**0.08**

4. Melting-Freeze Water Cycle. Formation of PAHs, Pyrimidenes and Triazines

Polycyclic aromatic hydrocarbons (PAHs) were proposed as key molecules in the study of the origin of life. Due to their photochemical properties, PAHs could have played the role of a primitive pigment system that drove the synthesis of amphiphilic compounds and contributed to prebiotic chemical evolution (Deamer 1992 and Segré et al. 1988). The "aromatic world" hypothesis suggests that assemblies based on aromatic hydrocarbons may have served as components of informational polymers, or as containers and mediators in protometabolic pathways (Ehrenfreund et al. 2006). The stability of such aromatic compounds and their capacity for self-assembly, driven by the π-π stacking interaction and weak forces, could make aromatic hydrocarbons potential building blocks for protocellular structures. Indeed, it has been demonstrated that amphiphilic polycyclic aromatic compounds are capable of self-assembly and may form bilayer structures (Chen et al. 1999). However, the formation of PAHs in prebiotic experiments has not been widely reported and only few examples have indicated their direct formation. They were identified by inferring data of L^2 MS spectroscopy (Imanaka et al. 2004 and Sagan et al. 1993).

As low temperatures favor the production of acetylene (main precursor of PAHs) and in order to explore the possible formations of PAHs in an icy environment, water freeze-melt cycles were simulated by using a freeze-melt cycle that was generated by the device shown in figure 5. Once the freeze-melt cycle was established the system was energized by spark discharges. A tungsten electrode connected to a high-frequency and high-voltage generator induced a discharge that impacted the water-ice interface. After 72h, the voltage generator was switched off and the reactor was maintained under active freeze-melt cycling.

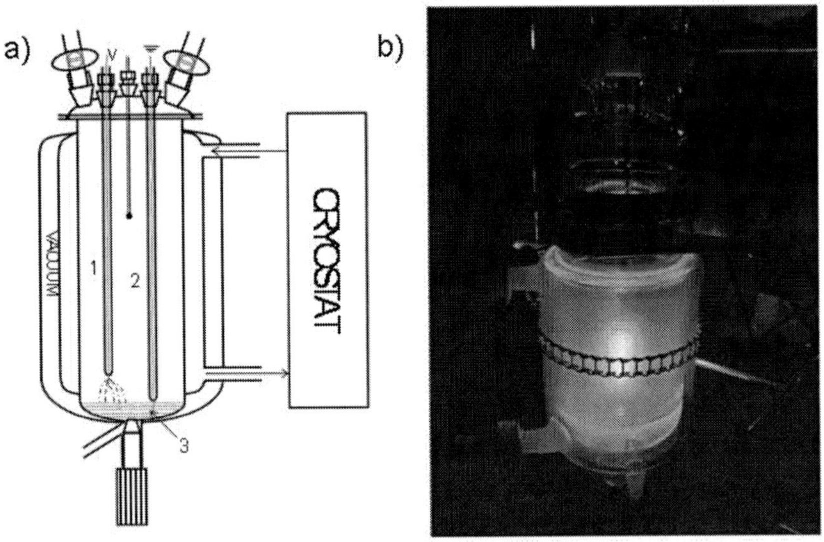

Figure 5. a) Device to generate freeze-melt cycles and to produce spark discharges on water solutions. 1, 2: Tungsten electrodes connected to a high-voltage (50 kV) generator and to the ground. 3: Pure water or water solution. (Menor-Salván et al. 2009). b) Photograph shown the freeze-melt cycles device using a UV lamp.

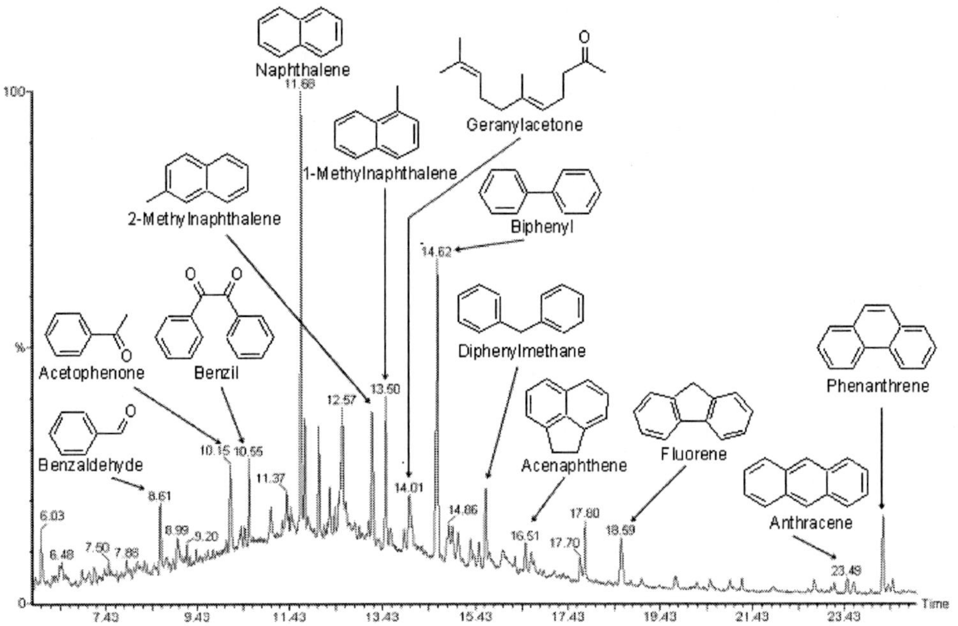

Figure 6. GC-MS total ion chromatogram of SPME extract of bulk sample obtained after 72 of energization by spark discharges and left stand for three months maintaining a water freeze-melt cycle under a initial $CH_4/N_2/H_2$ atmosphere.

Using the apparatus indicated above to simulate freeze-melt cycles, energizing a sterile water pool and $CH_4/N_2/H_2$ atmosphere during three days and maintaining an active freeze-melt cycle (between -5 ºC and 5 ºC) for three months, a yellowish solution with suspended organic solid was formed (Menor-Salván et al. 2008). The bulk sample was analyzed for PAHs by solid phase microextraction (SPME) coupled with gas chromatography/mass spectrometry (GC/MS) (Figure 6). A rich mixture of benzene derivatives with a significant abundance of phenylethyne and biphenyl compounds was found. We identified significant amounts of naphthalene, alkyl-naphthalenes, acenaphthene, fluorene and anthracene. Diphenylmethane, acenaphthylene, pyrene, anthracene, phenanthrene and benzyl, and the single aromatic rings, benzene and toluene, were also identified. The main aromatics synthesized were naphthalene, methyl-naphthtalenes and biphenyl, containing 0.01% of the total carbon introduced in the experiment. Approximately 0.31 μmL of PAHs were formed (the total amount of carbon introduced in the system was 428 mg). Moreover, the organic speciation indicates that some aromatic carbonyl derivatives (acetophenone, benzaldehyde, tolualdehydes, benzoic acid and geranylacetone) were also synthesized.

After analysis of the bulk sample, we performed the analysis of the suspended solid using APCI-TOF-MS. The total mass of the sample was 22.5 mg, representing 29% of the total non-volatile organic material obtained in the experiment and 2.6% of the total carbon introduced. This solid is partially soluble in water, and if left off stand, coalesces into micron-sized particles that absorb dyes and can be separated by centrifugation. The APCI fragmentation showed a polymeric nature for the solid, displaying several significant fragments consisting of successive losses of triacetylene (C_6H_2, m/z = 74) and water (m/z = 18). The polymer appears to be comprised by chains of repetitive units of hydroxyl substituted triacetylene with amine and nitrile terminations. No evidence of aromatic carbon was found in polymers or solids formed during the experiment and the largest PAHs found were three ring species. The proposed structure for this polymeric material is really different from the hydrophilic and hydrophobic *tholins* formed in the experiments with water aerosols and liquid water, as well as form the organic compounds identified. This highlights the importance of the temperature and the interfaces in prebiotic chemical processes.

To check the real influence of the freeze-melt cycles on the production of PAHs, four control experiments were performed using the same reactor: i) Same atmosphere and same reaction times with water in the liquid phase at room temperature. In this experiment, we did not find evidence of the synthesis of aromatic compounds, and aliphatic alcohols were identified in the reaction mixture: 3-methyl-butan-2-ol, pentan-2-one, pentan-2-ol, 3-methyl-pentan-2-ol, 3-methyl-pentan-1-ol, and 2-methyl-hexan-2-ol. This results suggests that a T > 5°C increases the atmospheric generation of $HO^{\cdot}$ radical that leads to the generation of alcohols and further oxidation to carbonyl compounds. ii) Same conditions but without energization of the system. In this case, neither organic material nor contaminations were observed in the water pool. iii) Same conditions as the ones for the freeze-melt cycles and spark discharges for 72 hours, but the samples were analyzed immediately after this period. We found substituted benzene rings and naphthalene derivatives: toluene, phenylethyne, styrene, benzaldehyde, benzonitrile, acetophenone, benzylcyanide, deca-4,6-diyne, 1-phenyl-1-propanone, naphthalene, 2-methylnaphthalene, and 1-methyl-naphtalene. iv) Same conditions as the main experiment in terms of time and energy and water in the ice phase without the melting protocol. Significant amounts of conjugated dienes, diynes and nitriles, i.e., hexa-2,4-diyne, 1-cyanobutadiene, hex-4-yn-3-one, phenylethyne, hexa-2,4-dienal, p-tolualdehyde, deca-4,6-diyne, cinnamonitrile, biphenyl, were identified but no evidence of PAHs synthesis. The formation of solid material with low solubility was increased in this control experiment. In this experiment, unsaturated molecules containing C=C or C≡C bonds and reactive groups (aldehyde, nitrile) may be precursors that undergo cyclization reactions to form more stable compounds such as PAHs.

Therefore, the freeze-melt cycles (that, from an astrobiological point of view, may be generated with water ice and liquid phases in equilibrium by thermal variations on a planetary surface) convert the ice surface from a trap or condenser of reactive molecules to a reactor that leads to the synthesis of a complex mixture of aromatic compounds and unsaturated molecules. This particular role of ice could be due to the generation of concentrated brines during freezing that are subjected to high pressure and solution gradients, and/or to catalytic effects of the ice surface. In this "ice reactor", reactions that lead to more stable and complex products, such as PAHs, and to the formation of carbonyl compounds may occur. Additionally, they may prevent the formation of soot or kerogen-type material by the generation of a hydrophilic poly(triacetylene) material that could form particles and vesicles by coalescence. Polyynes and cyanopolyynes have been detected in some astrophysical objects and the presence of triacetylene and more complex polyynes is expected in Titan (Cataldo 2006 and Aflalaye et al. 1995). These compounds are stable at low temperature and evolve to polymer chains at standard temperature. In our conditions, the generation of acetylenic reactants in the presence of ice leads to the formation of a OH-rich, polar poly(triacetylene) derivative. This result is interesting for the expected chemistry in possible surface or subsurface ices on planetological objects.

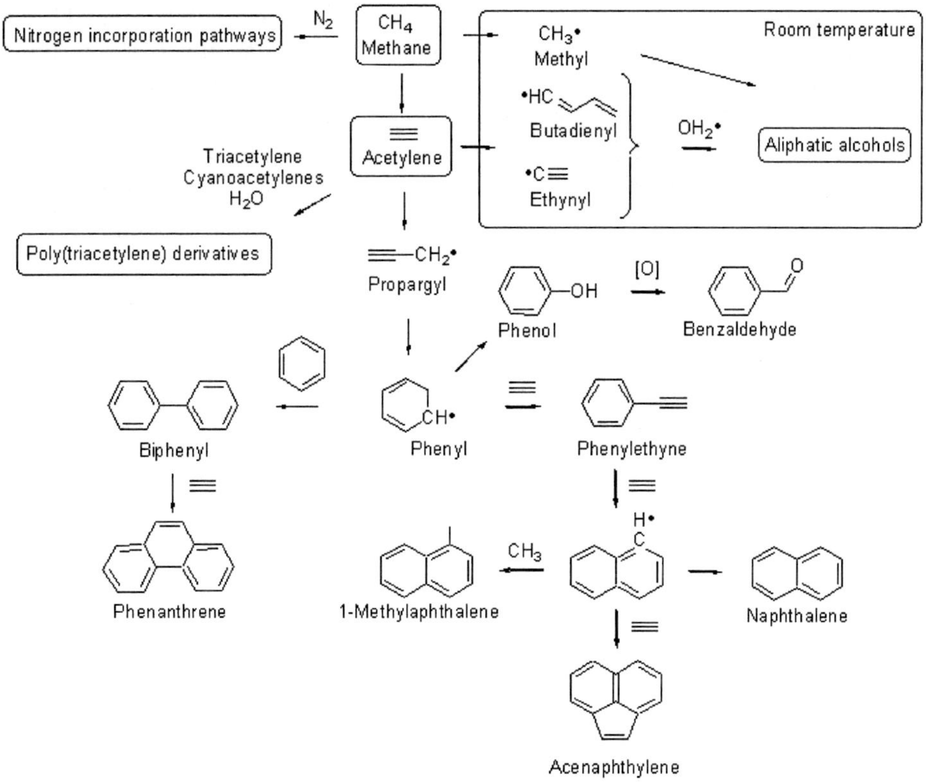

Figure 7. Possible pathway of reaction in freeze-melt experiments from methane. The boxed zone of pathway represents alternative route favored at room temperature with formation of alcohols. Low temperature favors the mechanism based on acetylene polymerization that generates PAHs and the poly(triacetylene) polymers.

As a summary of the effects of the water freeze-melt cycles, we can say that: i) Icy surfaces favor the synthesis of molecules with conjugated double/triple bonds and the freezing process favors cyclization processes and polar aromatic compounds. ii) Aromatic compounds may be synthesized in planetary/asteroidal environments, including terrestrial planets or Saturn's icy moons, from methane by the acetylene polymerization mechanism. iii) Irradiation of methane-rich ices or water ice under methane-nitrogen atmosphere generates a poly(triacetylene) derived *tholin* and discrete aromatic molecules, but not polyaromatic kerogen-type material or HCN polymers (Mamajanov and Herzfeld 2009, Matthews and Minard 2006).

If the most interesting finding in the water aerosols experiments was the formation of the carboxylic acids from OH· radicals and probably aliphatic radicals, in the case of the freeze-melt experiments it is the fact that acetylene plays the main role in the formation of PAHs, according to the scheme shown in figure 7.

The prebiotic synthesis of nucleic acid bases is a central issue in the RNA world hypothesis, one of the main proposals for the origin of life, based on the self-assembly of nucleic acid monomers. The possible scenarios for the synthesis of nucleic acids are still under debate, and despite the abiotic synthesis of several nucleobases, the relevance of these syntheses to the origin of life is not well established. In a pioneering work, Ferris, Sanchez and Orgel obtained cytosine from urea and cyanoacetylene and from a sodium cyanate solution and cyanoacetylene (Ferris et al. 1968). The isolation of the latter from spark discharges in methane-nitrogen mixtures suggests the relevance of this synthesis for the origin of life (Ferris et al. 1968, Sanchez et al. 1966). Robertson and Miller (1995) have reported on the synthesis of cytosine with high yields (30-50%) by heating various concentrations of urea and cyanoacetaldehyde in a sealed ampoule. Recently, the prebiotic relevance of pyrimidine synthesis from urea and cyanoacetaldehyde was discussed (Shapiro 2002 and 1999). In order to explore the prebiotic importance of the suggested cyanoacetylene/urea pathway to cytosine and to provide an icy scenario for the synthesis of pyrimidines and triazines from methane/nitrogen and urea, we used the same freeze-melt cycles as described above, i.e., a water solution of urea (0.1 M) was energized with spark discharges under a $CH_4/N_2/H_2$ atmosphere during three days. As in the previous case, the freeze-melt cycle was maintained between -5 °C and 5 °C for three weeks and a yellowish solution without any suspended organic solid was collected. The solution was freeze-dried and the dry yellow solid was analyzed for bases as their trimethylsilyl derivatives by GC-MS (Menor-Salván et al. 2009).

The reaction product (Figure 8) contained unreacted urea, triazine cyanuric acid (2,4,6-trihydroxy-s-triazine), ammelide (6-amino-2,4-dihydroxy-s-triazine), ammeline (4,6-diamino-triazin-2-ol), and melamine (2,4,6-triamino-s-triazine). The major products were cyanuric acid and cytosine (with a yield of 7.1 and 4.2 %, respectively, based on the total amount of urea introduced into the reactor). The only amino acids identified in the reaction product were glycine (not shown in the chromatogram), asparagine, and aspartic acid. The hydantoins were also present in minor quantities.

As in the case of the formation of PAHs, four control experiments were carried out to determine the real effect of the freeze-melt cycle on the formation of pyrimidines and triazines: i) A control experiment was performed at room temperature using the same experimental set-up. Under these conditions, the reaction product contained a significant amount of insoluble organic material (32.5% of the carbon introduced as methane and urea in the experiment). The solution contained unreacted urea, glycine, hydantoin, 5-hydroxy hydantoin, alloxanoic acid, parabanic acid, and carbamoyl-glycine. The generation of pyrimidines and purines was undetectable and triazines were detected in low yields (<0.01%). ii) An experiment was performed by sparking the urea solution under the same conditions as the freeze-melt cycles, but under an inert atmosphere (argon). In these conditions, we observed the formation of triazines, cyanuric acid, ammelide, and melamine, but not of pyrimidines. The yields of triazines obtained in this experiment are significantly lower than under a methane/nitrogen atmosphere. This could be due to the ionization of argon that diverts the electric discharge in all directions around the electrode, thus diminishing the energy input in the ice water pool. It is noteworthy that melamine, but not ammeline, was found in these conditions. iii) An experiment was performed at low temperature (-5° C) under the same conditions as the main experiment, but without freeze-melt cycles. Under these conditions, we found that triazine derivatives were mainly formed. We also found that the concentration of pyrimidines and the overall yields were significantly lower than the ones found for freeze-melt cycle experiments. iv) An experiment was performed without spark discharges in the $CH_4/N_2/H_2$ atmosphere. After 3 weeks of freeze-thaw cycles, the solution content was analyzed and no evidence of degradation, microbial contamination, or generation of nitrogen-containing heterocycles was found.

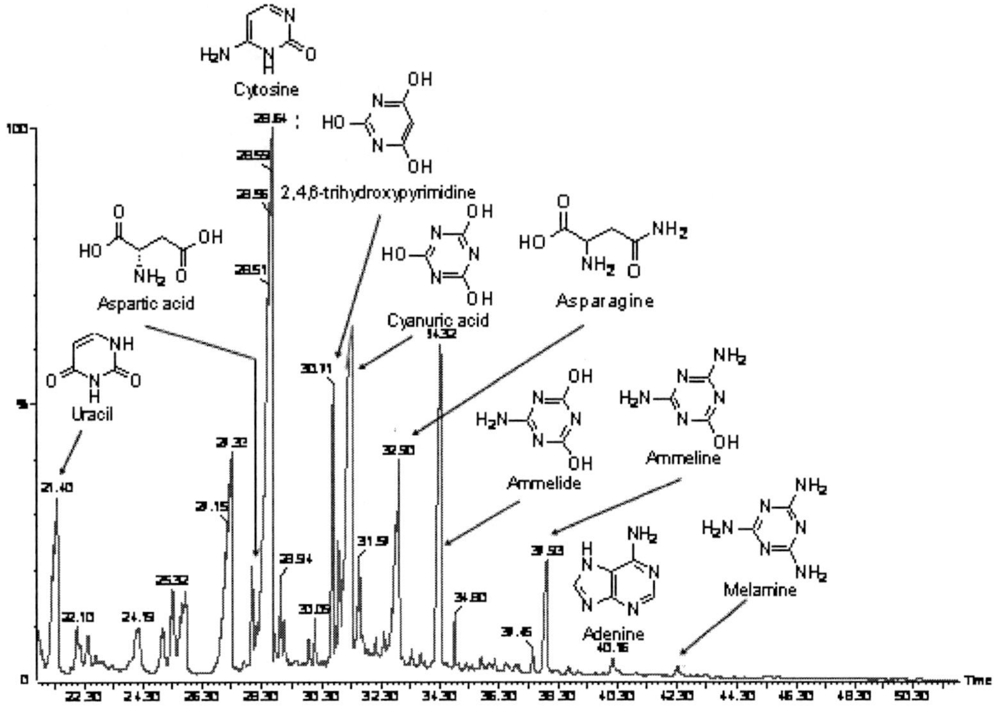

Figure 8. Total ion gas chromatogram showing the trimethylsilyl derivatives of bases, pyrimidines and trizines, found from the experiment of a water solution of urea (0.1 M) energized with spark discharges under a reductive atmosphere of $CH_4/N_2/H_2$ and subjected to a freeze-melt cycles.

Taking into account the results of the different experiments with the urea solution, it seems that an urea solution subjected to freeze-melt cycles favors the generation of nucleobases, diminishes the formation of insoluble *tholins* and other polymeric organic material, and reduces the importance of side reactions between urea and amino acids or their precursors generated in classic spark experiments. Moreover, pyrimidines are more stable at low temperatures and the ice matrix allows their chemical evolution to form more complex molecules. The microenvironments where the molecules are located are surrounded by a medium that is more reflective and can protect from photodimerization. It also provides a higher stability against deamination due to lower temperatures that also prevent from thermal decomposition. Our results confirm the potential importance of ice in planetary environments, such as Mars, the icy moons of Saturn, and the primitive Earth, as previously suggested (Orgel 2002).

The main conclusion of the experiment with freeze-melt cycles and urea is that the classical synthesis of cytosine and uracil from cyanoacetilene/cyanoacetaldehyde and urea (Robertson and Miller 1995, Ferris et al. 1968) is only possible with a urea solution under a CH_4/N_2 atmosphere when the solution is subjected to freeze-melt cycles.

In summary, the spark discharges in a methane/nitrogen atmosphere and freeze-melt cycles operate by two basic mechanisms. The first mechanism consists in the generation and polymerization of acetylene that lead to aromatic compounds and acetylene-derived polymers, as described previously and the second one consists in the generation of cyanide derivatives, such as cyanoacetylene. Urea could act as a cyanoacetylene scavenger, favoring the formation of nitrogen derivatives (Figure 9). Therefore, the presence of urea reduces the availability of reactants for the generation of cyanide or acetylene polymers. This explains the lack of insoluble organic materials obtained in experiments using urea/ice. At room temperature, it is the classic mechanism of atmospheric generation of *tholins* and amino acid precursors that appears to prevail, leading to the formation of hydantoins without the formation of pyrimidines or triazines, despite the presence of excess urea in the solution.

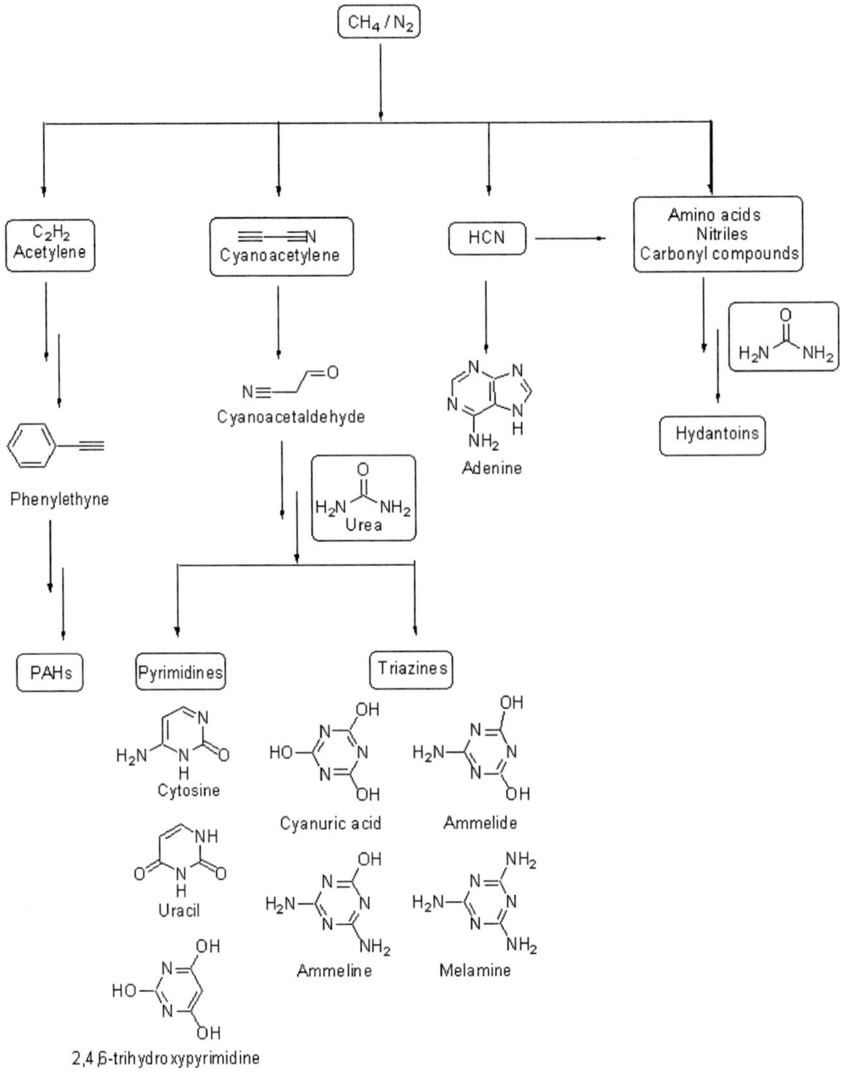

Figure 9. Possible synthetic pathways in a urea solution subjected to freeze-melt cycles under a CH_4/N_2 atmosphere, using spark discharges as an energy source. The pathways lead to the chemical families found in the reaction product: the synthesis of aromatics by the generation and polymerization of acetylene, the synthesis of pyrimidines from cyanoacetylene and urea, the synthesis of triazines from the possible isocyanic acid generated during urea decomposition and hydantoins generated from urea and amino acids, cyanhydrines, hydroxyl acids or other carbonylic precursors. These suggested pathways are based on the final products obtained. The pathway to hydantoins synthesis and insoluble organic material generation prevails in the liquid phase. Instead, frozen medium favours the generation of pyrimidines and triazines.

Additionally, the high-yield obtained in the production of triazines in freeze-melt cycles opens a new way to explore these molecules as mimics of prebiological nucleobases in the early stages of chemical evolution.

5. WATER INTERFACIAL PROCESSES IN CRYSTALIZATIONS SHOWING CHIRAL SYMMETRY BREAKING

Scientists look for processes that may have been responsible for the chiral symmetry breaking. The tiny excess of one enantiomer in a racemic mixture either due to natural fluctuations, external fields, or small energy imbalances must have been amplified by the joint effect of an autocatalytic process, i.e., some kind of competition between the left- and right-handed products, and the action of external agents that acted as chiral catalysts.

The interaction of chiral or achiral compounds with inorganic chiral crystalline surfaces to achieve chiral recognition and enantioselectivity is a frontier topic in chemistry (Hazen and Sholl 2003) with many experimental reports (see e.g., Attard et al. 2002). Furthermore, in autocatalytic reactions showing spontaneous symmetry breaking in a complex system displaying a Frank-like behavior, as it seems to be the case of the Soai reaction (Crusats et al. 2009), the chiral transfer from the chiral solid surface to the reacting species in the solution has been demonstrated (Kadowaki et al 2008; Kawasaki et al 2006; Soai et al 1999).

Chiral surfaces arise in chiral crystals (e.g., quartz), but intrinsically chiral surfaces can also appear in achiral crystals (e.g., calcite). The role of mineral chiral surfaces in the emergence of chirality in prebiotic scenarios has already been proposed (Hazen 2006; Lahav 1999). However, "although some of these chiral surfaces can serve as effective asymmetric inductors, however, ...it is unclear how chiral catalytic activity in a given sense could prevail over its equally probably mirror image at the same time in a different scenario." [p. 132 in (Guijarro and Yus 2009)].

Crystallizations processes of achiral compounds yielding enantiomorphic crystals are used as models to understand the basic principles acting on the spontaneous emergence of homochirality in autocatalytic processes and the robustness of biological homochirality. Achiral compounds of low-symmetry may act as chiral templates for other species in water solutions (Weissbuch et al. 2008). In a pioneering work, Lahav and Leiserowitz have shown that the crystalization of achiral glycine in a water/air interface may lead to the spontaneous deracemization of a racemic mixture of other chiral amino acids present in the solution; one enantiomer growths as crystals on the surface and the other enantiomer remains in solution (Weissbuch et al. 1997; Weissbuch et al. 2005).

$NaClO_3$ crystallization is a typical example of an achiral compound, or fast racemizing compound in solution, that yields enantiomorphic crystals as the most stable crystal mesoforms. This type of systems is sensitive to chiral inductions originated by certain chiral contaminants so that, instead of the racemic mixture, a final scalemic crystal mixture may be obtained. Moreover, there are reports that clearly show that in specific experimental conditions the system is so sensitive to chiral indutions that homochiral crystal mixtures can be obtained (Kondepudi and Asakura 2001).

5.A. Symmetry Breaking Processes in the Water Aerosol Cycle

In order to explain the final homochiral scenario of terrestrial life, it is necessary to investigate the role of all possible physical or chemical mechanisms of enantiomeric amplification. On Earth, the natural water cycle goes through the aerosol phase and the liquid bulk phase on a regular basis. Recently, the relevance of aqueous interfaces and the water aerosol cycle for the enhancement of an initial enantiomeric excess and for the spatial homogenization of the handedness of the final product has been shown experimentally. In particular, it has been demonstrated that an aerosol-liquid cycle of a $NaClO_3$ solution launches an autocatalytic amplification of an initial imbalance of about 10^{-7} % (1 ppb) to a total chiral purity in a one-pot, single-step process (Osuna-Esteban et al. 2008). In this experimental set-up, each aerosol was a micron-sized reactor that produces chiral nano-crystals of the same handedness (Figure 10). The aerosol crystallization process takes place on the droplet due to the intense evaporation, which is the first environment to reach supersaturation. The enantiomeric yield is then enhanced by the recycling of the crystallites through the liquid phase, where the smaller crystallites are dissolved. This method induces, even if it is in minute amounts, the homochiral crystallization of sodium chlorate for all the crystals produced within a reaction basin. Due to the accumulation at aerosols interfaces of ambient chiral impurities or added hydrophobic chiral amino acids in tiny concentrations (ppb), the initial levorotatory (l) and dextrorotatory (d) excess will not be produced with equal probability. In the absence of added catalysts, this process leads to preferential (d) homochiral crystallizations in a ratio of 4:1 that is due to ambient contamination. By adding only 2 ppb of an hydrophobic amino acid, such as (L) or (D) Phe, one is able to induce a final preferential homochiral crystallization of (d) or (l) handedness in a ratio of 2:1, respectively.

Figure 10. Experimental set-up to investigate the role of the water aerosol cycle and the aqueous interfaces on chiral symmetry breaking.

Figure 11. Scanning Electron Microscope (SEM) image of chiral sodium chlorate crystals formed during the intense evaporation process on the surface of aerosols.

A theory has suggested that discrimination between hetero- and homochiral orderings becomes more likely when short-range repulsive forces (such as those experienced by molecules in close-packed crystals) are significant. On the surface, short-range repulsive forces induce the production of homochiral crystal aggregates. Any initial bias (ambient chiral impurities or added hydrophobic chiral amino acids in concentrations of ppb) that may support the faster crystallization of one of the enantiomorphs in the aerosol air-water interface will make the final homochiral configuration more probable. As far as the influence of amino acids is concerned, hydrophilic amino acids are primarily solubilized in the water aerosol volume, whereas hydrophobic molecules migrate to the air-water interface (Yano et al. 2000). In this experiment, hydrophobic molecules located at the air-water interface are more effectively driving the chirality of the crystallization process than hydrophilic molecules (Figure 11).

5.B. Boiling Solutions

Recent results on the crystallization of systems composed of two components (achiral compound or chiral compound rapidly racemizing in the solution plus solvent) that yield chiral enantiomorphic crystals indicate that, under certain experimental conditions, only one of the solid enantiomorphs is obtained (Viedma 2005). Its chirallity is randomly distributed between several experiments. This implies that in some conditions the racemic state is a metastable state that leads to homochiral crystal mixtures of stochastic chiral sign. In such a system, the effect of tiny chiral inductions drives the system towards a specific chirality. This explains the strong chiral bias generated by small amounts of chiral contaminants in the aerosol crystallization discussed in the former section. The emergence of a bifurcation scenario leading to homochirality in such systems has been proposed. It would be due to the recycling through the primary nucleation stage, implying chiral recognition in the aggregation processes (Ribo et al. 2006). A simple explanation has been proposed on the basis of the thermodynamic differences of the systems when chiral recognition does and does not occur between the chiral solid phases. The experimental conditions that convert the racemic state in a metastable state are those of a wet mechanical stirring of crystal mixtures in equilibrium with their saturated solutions (Viedma 2005). However, it has been very recently reported that such a bifurcation scenario can also be obtained when the crystallization of achiral sodium chlorate $NaClO_3$ is performed in boiling solutions (El-Hachemi et al. 2009), corresponding to a possible terrestrial scenario of hot springs conditions. The experiment consisted in the crystallization from supersaturated boiling aqueous solution that is induced by withdrawal of the solvent from the reflux flow in the absence of seeding. The objective of these experiments was to assess the chirality of the crystal mixture nucleated in the crystal free solution, for that it was used a very low rate of solvent extraction. The low ratio of water/$NaClO_3$ at the detection of visible crystals, the sudden formation of a thick crystal solution at all the walls of the flask, and the fast filtration process provided a reasonable explanation for the chirality of the crystal, namely that the outcome originated from the equilibration during the first nucleation. The results on the chirality of the crystal powder mixtures, as determined from the ORD spectra (crystal powder dispersion in Nujol), are shown in Fig. 12 for a total of 32 different crystallizations (no experiment was discarded): 17 were levorotatory, 13 were dextrorotatory, and only 2 experiments yielded a non-optically active outcome. Therefore, the results strongly suggest that the pseudo-equilibrium at the primary nucleation "counter-intituitively" corresponds to a homochiral state or a scalemic mixture indistinguishable from homochirality within the experimental error. A peculiarity of crystal growth in boiling solutions is the formation of a temperature gradient between the liquid surface and the bottom of the reactor. This is caused by the superheating required by the nucleation of bubbles. Therefore, this temperature inhomogenity and the one originated by the arrangement of the solution species in the water/air interface are probably the reasons why the system is far from equilibrium and the racemic state is converted in a metastable state.

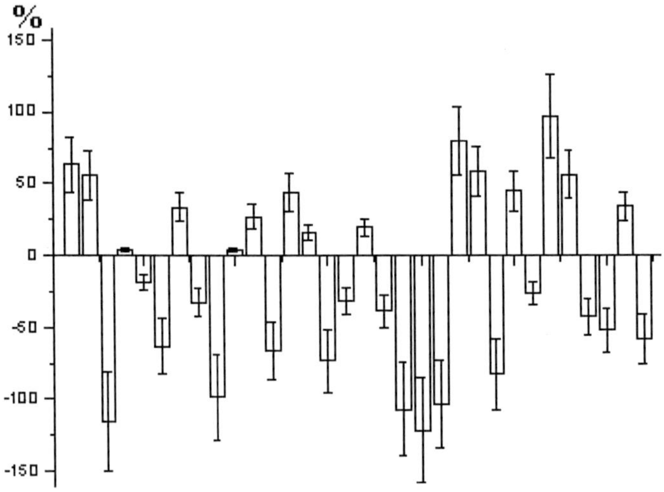

Figure 12. Percentage of the CB values (mdeg mg^{-1} at 235 from the ORD spectra) of the crystal powder outcome of the $NaClO_3$ boiling crystallization with respect to the corresponding values of pure d- and l-crystals.

6. CONCLUSIONS

Classic prebiotic chemistry experiments only consider a single water state (gaseous, liquid or solid). Herein, we have summarized experimental studies that demonstrate the importance of interfaces between different water phases (such as liquid-gaseous, or liquid-ice) that can have a positive influence on the reactivity of the system despite other constant environmental variables. We have observed that, under a variety of experimental conditions, the existence of water interfaces enhances the amount and diversity of organic compounds obtained as products of the reacting system.

It is also worth mentioning that, under similar environmental conditions and energy sources, different states of water interfaces lead to different products. In particular, the main products in aerosol vapor-liquid experiments were carboxylic acids, whereas the most relevant ones in ice-liquid experiments were polycyclic aromatic hydrocarbons.

Interfaces change the mobility of reactants, favor their concentration and local ordered arrangement of molecules. This amy have had an influence on the amplification of the chiral sign of the obtained products.

Furthermore, water interfaces that are subjected to phase changes create local dynamic environments where the pH (free hydroxylic radicals), enthalpy, hydration, and dehydration levels may change continuosly. These changes may active the reactants molecules and modulate their reactivity.

7. REFERENCES

Aflalaye, A; Andrieux, D; Bénilan, Y; Bruston, P; Coll, P; Coscia, D; Gazeau, MC; Khlifi, M; Paillous, P; Sternberg, R; de Vanssay, E; Guillemin, JC; Raulin, F. *Adv. Space Res.*, 1995, 15, 5-11.

Attard, GA; Harris, C; Herrero, E; Feliu, J. *Faraday Discuss*, 2002, 121, 253-266.

Bada, JL; Bigham, C; Miller SL. *Proc. Natl. Acad. Sci., USA*, 1994, 91, 1248-1250.

Barley, ME; Pickard, AL; Sylvester, P. *J. Nature*, 1997, 385, 55-58.

Bernstein, MP; Dworkin, JP; Sandford, SA; Cooper, GW; Allamandola, L. J. *Nature*, 2002, 416, 401-403.

Bibring, JP. et al. *Nature*, 2004, 428, 627-630.

Borquez, E; Cleaves, HJ; Lazcano, A; Miller, SL. *Orig. Life Evol. Biosph.,* 2005, 35, 79-84.

Boynton, WV. and 24 collegues. *Science*, 2002, 297, 81-85.

Buser, HJ; Schwarzenbach, D; Petter, W; Ludi, A. *Inorg. Chem.*, 1977, 16, 2704-2710.

Cataldo, F. *Orig Life Evol Biosph.*, 2006, 36, 467-475.

Chen, L; Geiger, C; Perlstein, J; Whitten, DG. *J. Phys. Chem., B*, 1999, 103, 9161-9167.

Crusats, J; Hochberg, D; Moyano, A; Ribo, JM. *Chemphyschem*, 2009, in press.

Deamer, DW. *Adv. Space. Res.,* 1992, 12, 183-189.

Donaldson, DJ; Vaida, V. *Chem. Rev.*, 2006, 106, 1445-1461.

Donaldson, DJ; Tervahattu, H; Tuck, AF., Vaida, V. *Orig. Life Evol. Biosph.,* 2004, 34, 57-67.

Dobson, CM; Ellison, GB; Tuck, AF., Vaida, V. *Proc. Nat. Acad. Sci., USA*, 2000, 97, 11864-11868.

Donaldson, DJ; Anderson, D. *J. Phys. Chem., A*, 1999, 103, 871-876.
Ehrenfreund, P; Rasmussen, S; Cleaves, J; Chen, L. *Astrobiology*, 2006, 6, 490-520.
El-Hachemi, Z; Crusat, J; Ribo, JM; Veintemillas-Verdaguer S. *Cryst. Growth Design*, 2009, submitted.
Ellison, GB; Tuck, AF; Vaida, V. *J. Geophys. Res.*, 1999, 104, 11633-11641.
Encrenaz, T; Melchiorri, R; Fouchet, T; Drossart, P; Lellouch, E; Gondet, B; Bibring, JP; Langevin, Y; Titov, D; Ignatiev, N; Forget, F. *Astronomy and astrophysics*, 2005, 441, L9-L12.
Fanale, FP; Salvail, JR; Bruce Banerdt, W; Saunders, RS. *Icarus*, 1982, 50, 381-407.
Feldman, WC; Boynton, WV; Tokar, RL; Prettyman, TH; Gasnault, O; Squyres, SW; Elphic, RC; Lawrence, DJ; Lawson, SL; Maurice, S; McKinney, GW; Moore, KR; Reedy, RC. *Science*, 2002, 297, 75-78.
Ferris, JP; Sanchez, RA; Orgel, LE. *J. Mol. Evol.*, 1968, 33, 693-704.
Fouchet, T; Lellouch, E; Ignatiev, NI; Forget, F; Titov, DV; Tschimmel, M; Montmessin, F; Formisano, V; Giuranna, M; Maturilli, A; Encrenaz, T. *Icarus*, 2007, 190, 32-49.
Garrison, WM; Morrison, DC; Hamilton, JG; Benson, AA; Calvin, M. *Science*, 1951, 114, 416-417.
Groth, W; Suess, H. *Naturwissenschaften*, 1938, 26, 77-77.
Grotzinger, JP; Kasting, JFJ. *Geol*, 1993, 101, 235-243.
Guijarro, A; Yus, M. *The origin of chirality in the molecules of life*. RSC Publishing, Cambridge, UK, 2009.
Hazen, RM. Am. *Mineralogist*, 2006, 91, 1715-1729.
Hazen, RM; Sholl, DS. *Nature Mat*, 2003, 2, 367-374.
Holland, HD. The chemical evolution of the atmosphere and oceans, Princeton University Press, Princeton, *NJ*, 1989, Chapter 8.
Imanaka, H; Khare, BN; Elsila, JE; Bakes, ELO; McKay, CP; Cruikshank, DP; Sugita, S; Matsui, T; Zare, RN. *Icarus*, 2004, 168, 344-366.
Jortner, *J. Phil. Trans. R. Soc. B*, 2006, 361, 1877-1891.
Kadowaki, K; Suzuki, K; Hakoda, Y; Soai, K. *Angew. Chem. Int. Ed.*, 2008, 47, 496-499.
Kargel, JS. *Science*, 2006, 311, 1389-1391.
Kasting, JF. Evolution of the Earth´s atmosphere and hydrosphere: Hadean to recent. In MH; Engel, SA. Macko, (Eds.) Organic Geochemistry: Principles and Applications, Plenum Press, New York, 1993, 611-623.
Kawasaki, T; Suzuki, K; Hatase, K; Otsuka, M; Koshima, H; Soai, K. *Chem. Comm.*, 2006, 17, 1869-1871.
Kempe, S; Kazmierczak, J. *Astrobiology*, 2002, 2, 123-130.
Kieffer, HH; Titus, TN. TES mapping of Mars' north seasonal cap. *Icarus*, 2001, 154, 162-180.
King, MD; Thompson, KC; Ward, AD. *J. Am. Chem. Soc.*, 2004, 126, 16710-16711.
Kivelson, MG; Khurana, KK; Russell, CT; Volwerk, RJ; Walker, RJ; Zimmer, C. *Science*, 2000, 289, 1340-1343.
Knauth, LP. *Palageo. Palaeoclimat. Palaeoecol*, 2005, 219, 53-59.
Kobayashi, K; Masuda, H; Ushio, K; OACI, A; Yamanashi, H; Kaneko, T; Takahashi, J; Hosokawa, T; Hashimoto, H; Saito, T. *Adv. Space Res.*, 2001, 27, 207-215.
Kobayashi, K; Kasamatsu, T; Kaneko, T; Koike, J; Oshima, T; Saito, T; Yamamoto, T; Yanagawa, H. *Adv. Space Res.*, 1995, 16, 21-26.

Kobayashi, K; Tsuchiya, M; Oshima, T; Yanagawa, H. *Orig. Life Evol. Biosph.*, 1990, 20, 99-109.
Kondepudi, DK; Asakura, K. Acc. *Chem. Res.*, 2001, 34, 946-954.
Lahav, M. Biogenesis: *Theories of life's origin.* Oxford University Press, New York. 1999, 259.
Lerman, L; Teng, J. Ed. J. Seckbach. Springer. In the beginning. In Origins: Genesis and Diversity of Life, 2004, 6, 35-55.
Lerman, L. *Origins Life Evol. B*, 1996, 26, 369-370.
Levy, M; Miller, SL; Brinton, K; Bada, JL. *Icarus*, 2000, 145, 609-613.
Löeb, W. *Ber*, 1913, 46, 684-697.
Lorenz, RD; Stiles, BW; Kirk, RL; Allison, MD; Del Marmo, PP; Iess, L; Lunine, JI; Ostro, SJ; Hensley, S. *Science*, 2008, 319, 1649-1651.
Mahajan, TB; Elsila, JE; Deamer, DW; Zare, RN. *Orig. Life Evol. Biosph*, 2003, 33, 17-35.
Mamajanov, I; Herzfeld, J. *J. Chem. Phys.*, 2009, 130, 134503.
Manolache, S; Romanescu, C; Cobialeac, G; Simonescu, CI. *Molecules*, 1997, 2, 158-164.
Matthews, CN; Minard, RD. *Faraday Discuss*, 2006, 133, 393-401.
McDonald, GN; Khare, BN; Thompson, WR; Sagan, C. *Icarus*, 1991, 94, 354-367.
Macleod, G; Mckeown, C; Hall, AJ; Russell, MJ. *Orig. Life Evol. Biosph.*, 1994, 24, 19-41.
Meléndez-Hevia, E; Wadell, TG; Cascante, M. *J. Mol. Evol.*, 1996, 43, 293-303.
Mennella, V; Muñoz-Caro, GM; Ruiterkamp, R; Schutte, WA; Greenberg, JM; Brucato, JR; Colangeli, L. *Astron. Astrophys*, 2001, 367, 355-361.
Menor-Salván, C; Ruiz-Bermejo, M; Gúzman, MI; Osuna-Esteban, S; Muñoz-Caro, G; Veintemillas-Verdaguer, S. *Chem. Eur. J*, 2009, 15, 4411-4418.
Menor-Salván, C; Ruiz-Bermejo, M; Osuna-Esteban, S; Muñoz-Caro, G; Veintemillas-Verdaguer, S. *Chem. Biodiver.*, 2008, 5, 2729-2739.
Miller, SL. The formation of organic compounds on the primitive Earth. *Ann. N. Y. Acad. Sci.*, 1957, 69, 260-274.
Miller, SL. *J. Am. Chem. Soc.*, 1955, 77, 2351-2361.
Miller, SL. *Science*, 1953, 117, 528-529.
Miyakawa, S; Cleaves, HJ; Miller, SL. *Orig. Life Evol. Biosph*, 2002a, 32, 209-218.
Miyakawa, S; Yamanashi, H; Kobayashi, K; Cleaves, HJ; Miller, S. *Proc. Natl. Acad. Sci. USA*, 2002b, 99, 14628-14631.
Miyakawa, S; Tamura, H; Sawaoka, AB; Kobayashi, K. *Appl. Phys. Lett*, 1998, 72, 990-992.
Morse, JW; Mackenzie, FT. *Aquatic Geochem*, 1998, 4, 301-319.
Muñoz-Caro, GM; Meirhenrinch, UJ; Schutte, WA; Barbier, B; Arcones-Segovia, A; Rosenbauer, H; Thiemann, WHP; Brack, A; Greenberg, JM. *Nature*, 2002, 416, 403-406.
Muñoz-Caro, GM; Ruiterkamp, R; Schutte, WA; Greenberg, JM; Mennella, V. *Astron. Astrophys*, 2001, 367, 347-354.
Murphy, DM. *Science*, 2005, 307, 1888-1889.
Murphy, DM; Thomson, DS; Mahoney, MJ. *Science*, 1998, 282, 1664-1667.
Negron-Mendoza, A; Ponnamperuma, C. Orig. *Life*, 1976, 7, 191-196.
Orgel, L. *Orig. Life Evol. Biosph*, 2002, 32, 279-281.
Orgel, LE. *Proc. Nat. Acad. Sci. USA*, 2000, 97, 12503-12507.
Osuna-Esteban, S; Zorzano, MP; Menor-Salván, C; Ruiz-Bermejo, M; Veintemillas – Verdaguer, S. *Phys. Rew. Lett*, 2008, 100, 146102.
Plankensteiner, K; Reiner, H; Schranz, B; Rode, BM. *Molecular Diversity*, 2006, 10, 3-7.

Plankensteiner, K; Reiner, H; Schranz, B; Rode, BM. *Angew. Chem. Int. Ed.*, 2004, 43, 1886-1895.

Poulton, SW; Fralick, PW; Canfield, DE. *Nature*, 2004, 431, 173-177.

Robertson, M; Miller, SL. *Nature*, 1995, 375, 772-774.

Ribo, JM; Crusats, J; Veintemillas-Verdaguer, S. *Chem. Eur. J*, 2006, 12, 7776-7781.

Ruiz-Bermejo, M; Menor-Salván, C; Mateo-Martí, E; Osuna-Esteban, S; Martín-Gago, JA; Veintemillas-Verdaguer, S. *Icarus*, 2008, 198, 232-241.

Ruiz-Bermejo, M; Menor-Salván, C; Osuna-Esteban, S; Veintemillas-Verdaguer, S. *Orig. Life Evol. Biosph*, 2007, 37, 123-142.

Ruiz-Bermejo, M; Menor-Salván, C; Osuna-Esteban, S; Veintemillas-Verdaguer, S. *Orig. Life Evol. Biosph*, 2007b, 37, 507-521.

Russell, MJ; Hall, AJ. *J. Geol. Soc. Lond*, 1997, 154, 377-402.

Sagan, C; Khare, BN; Thompson, WR; McDonald, GD; Wing, MR; Bada, JL; Vo-Dinh, T; Arakawa, ET. *Astrophys. J*, 1993, 414, 399-405.

Sanchez, RA; Ferris, JO; Orgel, LE. *Science*, 1966, 154, 784-785.

Schlesinger, G; Miller, SL. *J. Mol. Evol.*, 1983, 19, 376-382.

Segré, D; Ben-Eli, D; Deamer, DW; Lancet, D. *Orig Life Evol Biosph*, 1988, 31, 57-63.

Shah, DO. In C. Ponnamperuma, (Ed.), chapter 7, *Exobiology*, North Holland, 1970, 235-265.

Shapiro, R. *Orig. Life Evol. Biosph*, 2002, 32, 275-278.

Shapiro, R. *Proc. Natl. Acad. Sci. USA*, 1999, 96, 4396-4401.

Sharpe, AG. *The chemistry of cyano complexes of transition metals*, London, Academic Press. 1976.

Simionescu, CR; Denes, F; Onac, D; Bloos, G. *Biopolymers*, 1974, 13, 943-954.

Smith, MD. *J. Geophys. Res.*, 2002, 107(E11), 5115.

Soai, K; Osanai, K; Kadowaki, K; Yonekubo, S; Shibata, T; Sato, I. *J. Am. Chem. Soc.*, 1999, 121, 11235-11236.

Steckel, H; Eskandar, F. Eur. *J. Pharm. Sci.*, 2003, 19, 443-455.

Stribling, R; Miller, S.L. *Origins of Life*, 1987, 17, 261-273.

Summers, DP. *Origins Life Evol. Biosph*, 2005, 35, 299-312.

Summers, DP; Lerner, N. *Origins Life Evol. Biosph*, 1998, 28, 1-11.

Taillades, J; Beuzelin, I; Garrel, L; Tabacik, V; Bied, C; Commeyras, A., *Orig. Life Evol. Biosph*, 1998, 28, 61-77.

Takahashi, J; Masuda, H; Kaneko, T; Kobayashi, K; Saito, T; Hosokawa, T. *J. App. Phys.*, 2005, 98, 024097.

Takano, Y; Ohashi, A; Kaneko, T; Kobayashi, K. *Appl. Phys. Lett*, 2003, 84, 1410-1412.

Tervahattu, H; Tuck, A; Vaida, V. Chemistry in prebiotic aerorsols: A mechanism for the origin of life. J. Seckbach, (Ed). Springer. In Origins: Genesis, Evolution and Diversity of Life, 2004, 6, 153-165.

Tervahattu, H; Juhanoja, J; Kupiainen, K. *J. Geophys. Res.*, 2002, 107, 10.1029/2002JD001403.

Trinks, H; Schröder, W; Biebricher, CK. *Orig. Life Evol. Biosph*, 2005, 35, 429-445.

Tuck, A. *Surv. Geophys*, 2002, 23, 379-409.

Urey, HC. *Proc. Natl. Acad. Sci. USA*, 1952, 38, 351-363.

Utsumi, Y; Hattori, T. *Review of Scientific Instruments*, 2002, 73, 1387-1389.

Viedma, C. *Phys. Rev. Lett*, 2005, 94, 065504(4).

Walker, JCG; Brimblecombe, P. *Precambian Res.*, 1985, 28, 205-222.

Weissbuch, I; Leiserowitz, L; Lahav, M. *Current Opinion in Colloid & Interface Science*, 2008, 13, 12-22.

Weissbuch, I; Leiserowitz, L; Lahav, M. *Top Curr Chem.*, 2005, 259, 123-156.

Weissbuch, I; Berfeld, M; Bouwman, W; Kjaer, K; AlsNielsen, J; Lahav, M; Leiserowitz, L. *J. Am. Chem. Soc.*, 1997, 119, 933-942.

Wilson, EH; Atreya, SK. *Plan. Spa. Sci.*, 2003, 51, 1017-1033.

Williams, RJP; Frausto da Silva, JJR. The natural selection of the chemical elements, *Claredon*, Oxford, 1996, 305 ff.

Yano, J; Furedi-Milhofer, H; Wachter E; Garti, N; *Langmuir*, 2000, 16, 9996.

Yuasa S; Flory, D; Basile, B; Oro, J. *J. Mol. Evol.*, 1984, 21, 76-80.

Zent, AP; Hecht, MH; Cobos, DR; Campbell, GS; Campbell, CS; Cardell, G; Foote, MC; Wood, SE; Mehta, M. *Journal of Geophysical Research-Planets*, 2009, 114, E00A27.

Zepik, H; Shavit, E; Tang, M; Jensen, TR; Kjaer, K; Bolbach, G; Leiserowitz, L; Weissbuch, I; Lahav M. *Science*, 2002, 295, 1266.

In: Astrobiology: Physical Origin, Biological Evolution... ISBN: 978-1-60741-290-8
Editors: S. Hegedus, J. Csonka, pp. 59-79

Chapter 3

INTEGRATION OF PRIMITIVE METABOLIC, INFORMATION AND STRUCTURAL PROTOCELL COMPONENTS UNDER SIMULATED EARLY EARTH CONDITIONS

***S. E. Maurer*[1,2] *and P. A. Monnard*[1,*]**

[1]FLinT Center Institute of Physics and Chemistry, University of Southern Denmark, Campusvej, 55, 5230 Odense M, Denmark

[2]Dept of Chemistry and Biochemistry, University of California, Santa Cruz, 1156 High Street, Santa Cruz, CA 94060

ABSTRACT

The first forms of living system, or protocells, must have been composed of chemical components or subsystems allowing them to ensure their maintenance, growth, and evolution. As such these subsystems must have possessed similar functions to those of present cellular life. However, contemporary cells rely on the interconnected work of information polymers, catalytic polymers, and their compartments, whose design required a long evolution.

To model plausible protocells and their functions, several classes of simpler compounds have been proposed and investigated, such as single chain amphiphiles (e.g., fatty acids) as building blocks of the compartments, polycyclic aromatic hydrocarbons (PAHs) as energy harvesting molecules, and catalytic RNAs (ribozymes) as both information and catalytic components. While very interesting insights in the processes and/or functions of each molecule class have already been gathered, it has become increasingly clear that the study of the protocell as a whole must be carried out to achieve the construction of a truly functional model system.

This review addresses the question of the formation of a protocell and its function at the system level. It will further deal with the environmental pressures likely present on the early Earth, as they would have influenced the lifetime of reactive components

[*] Corresponding author: Email: monnard@ifk.sdu.dk Phone +45 6550 4437, Fax: + (45) 6615 8760

needed for the start of life. The capabilities and limitations of protocell functionalities can be elucidated only through understanding the container interactions/interconnected functions with the other subsystems including environmental factors.

INTRODUCTION

Modern cells are defined relative to their external environment by a boundary, or membrane, that is composed of two complex layers formed predominantly by lipids, in which other functional components such as membrane proteins are embedded. This boundary, acting as a barrier to the free diffusion of solutes, selectively regulates exchanges of waste and nutrients between the environment and the cell. Membranes are also vital to the existence of a cell as means of sustaining a metabolic and genetic reproduction while preserving cellular contents. Their properties allow us to extract the minimal requirements for life.

Minimal life would have consisted of a metabolism, an information component (the precursor of genetic system), and a contained space (Figure 1). The building blocks (e.g. a nucleotide for an information polymer) of these three elements would all need to be produced by the metabolism in a linked fashion. This interconnection has lead to the hypothesis in which the informational component is also the metabolic component (RNA world[1, 2]), or that the container is composed of the same material as the metabolic component (proteinoid containers)[3]. Regardless of the association, these parts define cellular life. It should be noted that the ALL of these are essential for the existence of life. For example, a virus is often not considered alive because it lacks the metabolism needed to reproduce its genetic components or the proteins forming its capsids.

Clearly, it would be exciting to construct a self-reproducing chemical system. Such a system would represent a major step forward in understanding the origin of life, and in fact could help define what we mean by the living state. There would also be practical applications such as microscopic reaction vessels to replace bacteria in customized industrial scale synthesis of pharmaceuticals or petroleum. A further possibility, although it has mostly been a topic in science fiction, is that synthetic nanobots may soon become reality, just as evolutionary processes have produced the nanobots called antibodies. Another possible application is to study synthetic controllable “ecosystems”. However, our goal is to better define and model the conditions needed for life to arise in the first place.

The origin of life is a difficult research field due to the lack of firsthand information: Plate tectonics has erased most of the early Earth rock records and consequently evidence of the environmental conditions prevalent at that time have been lost. Scarce information sources, such as refractory zirconium-containing pellets formed between 4.0-4.5 Ga whose content could not be altered after their formation[4], can give us some information about the conditions prevailing at their creation[5]. This fact leaves researchers with only few hints to elucidate the events leading to the emergence of the first life forms, called protocells. Consequently, the scenarios proposed to describe the emergence of life have emphasized one essential aspect of life, usually directly related to researcher(s)’ field of expertise: The RNA world[1, 6], Lipid world[5], Aromatic world[7], or Metabolic world[8, 9] to cite the main hypotheses have been considered with a focus on information/catalysis, container, energy and metabolic cycles, respectively. While being important to strengthen our understanding of the properties of each component, this approach may well neglect an important aspect of life and by extension early forms of the emergent properties derived from the interacting components in systems.

Nonetheless, some general agreements have been reached. One is that complex organic molecules began to accumulate from general chemical reactions[10-12] and extraterrestrial sources[13, 14] when the Earth was starting to cool. The emergence of the protocell would have then occurred only by relying on self-assembly of chemicals into the protocellular components, for example, amphiphiles capable of self-assembly in solution could have helped to concentrate chemicals in structured or semi-structured aggregates. These systems could have represented the first containers that were available on early Earth. The exact set of the chemicals available at that time is still unclear and would have been highly dependent on the prevalent environmental conditions.

The lack of direct evidence about the conditions on the primitive Earth is clearly an issue when constructing a protocell chemical model. The presence of continents and liquid water, the best solvent for life, as far back as 4.3 Ga seems highly probable[4] as do higher volcanism and lower sun light irradiation compared to today. The cooling of the molten core likely resulted in higher surface temperatures. Because liquid water would first appear around 100°C and then slowly cool, it has been suggested that the temperatures where life first formed might have been higher than today's surface temperatures. In addition, areas of high volcanic activity and hydrothermal venting are suggested as prime reactors for the formation of high-energy molecules that could have significantly contributed to the formation of early metabolisms[15]. Another parameter that clearly influences synthetic reactions as well as self-assembly processes is the salinity of the solvent. Theoretical discussions of the ionic strength of early oceans have suggested an ocean twice as salty as it is currently[16]. Obviously, these environmental parameters must have significantly affected the processes leading to life's emergence, i.e. the formation of protocell, as well as the protocell chemical compositions and functions.

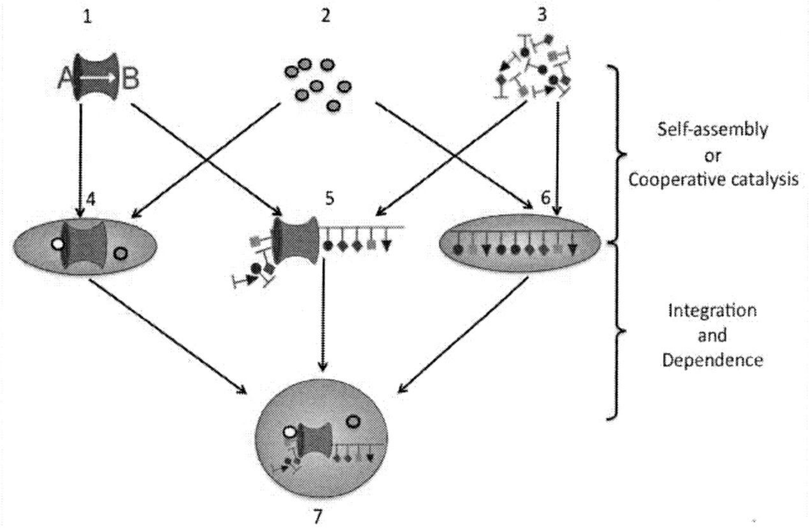

Figure 1. *Protocell component integration.* Metabolism (1), information (2), and containers (3) are necessary for the formation of the first life. The integration of two of these parts (4-6) has been the primary focus of most research to date. The future of the bottom-up approach to the origin of life question will be to integrate all 3 parts into a whole as their interconnections define emergent properties (7).

The focus of this review is to address the possible containers available on early Earth with respect to environmental factors. The interaction of these structures with other biomolecules that could have served in either information or metabolic assemblies will be of primary interest, with a brief discussion on primitive life cycles.

CONTAINER GENERATION

Any structure that restrained or confined a microenvironment/group of molecules could have been a prebiotic container. The ability to localize and protect a specific molecule, or set of molecules is considered an essential step for the formation of life.

One such structure occurs at inorganic mineral surfaces[17-21], where molecules are ordered and can diffuse on and off of the surface. This would allow for the concentration of molecules as well as their protection from reactive species. The ordering effect can also be used as a primitive enzyme to order molecules so that they have a better alignment for reaction, such as polymerization. Clays, quartz, and pyrite have all been suggested as mineral containers.

Aerosols have also been suggested as primitive compartment systems[22]. Average aerosol particles are about the same size as single-cell organisms and have the advantage of not being diluted by a salty ocean. Synthetic reactions could then occur in the upper atmosphere under intense UV light and at much cooler temperatures, making aerosol droplets desirable prebiotic containers for the production of biomolecules.

The most convincing and well-studied systems are amphiphilic systems (simpler than modern phospholipids) that are capable of self-assembly into cell-like membranes[23-25]. The conditions of early earth make it likely that amphiphilic hydrocarbons, such as fatty acids, were readily synthesized. This has been shown using Fischer-Tropsch-like reactions[26-28], where high pressure and a mixture of gases yield high hydrocarbon concentrations. Other fatty acid derivatives can then be synthesized, such as mono-acyl glycerols (GMAs)[29]. Polyprenyl compounds also show promise as a primitive amphiphile because of terpenoid universality in membranes across all classes of life as well as their non-biological plausability[30]. This leads to the conclusion that a wide variety of headgroups were likely under various prebiotic conditions to yield a complex mixture of non-biological amphiphiles (Figure 2).

Amphiphile molecules also accumulated during the heavy meteorite bombardment of early Earth through the late stages of accretion. As the solar system was newly formed, there was a large amount of cosmic debris, which was impacting the Earth in substantial numbers[31] delivering some organic matter. Dozens of carbonaceous chrondrite meteorites[32, 33] have been discovered and are more recent examples of such debris. These fragments were shown to contain organic material including polycyclic aromatic hydrocarbons (PAHs) and fatty acids (FAs). Extracts from the meteorites self-assembled into bilayers[34] as do organic compounds synthesized under modeled interstellar conditions[35], proving that mixtures of organic matter, synthesized in the interstellar environment have the capacity to form prebiotic containers.

The contribution of organic molecules by extraterrestrial sources has been estimated. The current quantity of interplanetary dust particles is 3×10^5 kg per year[36]. Early in the Earth's history it is estimated that 200 times as much organic material was deposited from space each year.

Nomenclature	Fatty acid (FA)	Fatty amine	Fatty Alcohol	Glycerol mono-acyl amphiphile (GMA)	Isoprenoids
Chemical structure	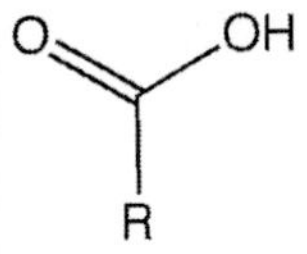	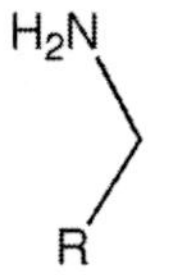		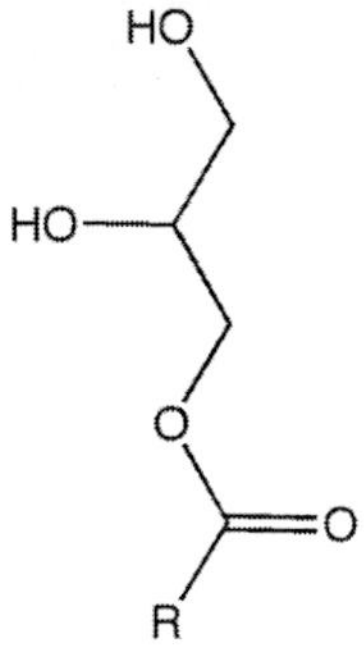	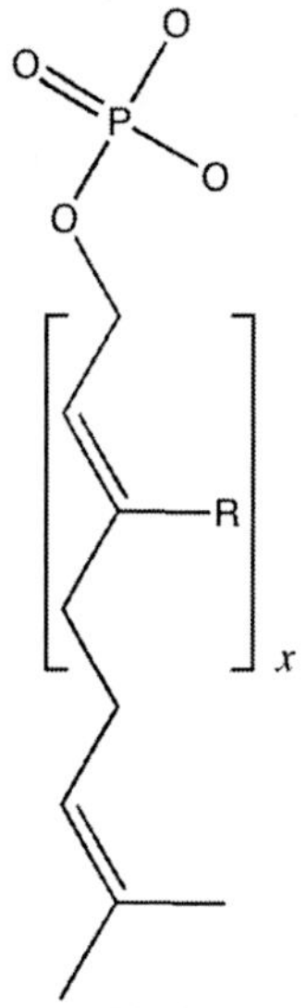
Formation restrictions	pH=pKa	pH≠pKa (when mixed with FA)	Co-surfactant only	Decreases salt and pH sensitivity	Unknown
Prebiotic relevance:	Meteorites, Prebiotic synthesis	Unknown	Meteorites, Prebiotic synthesis	Prebiotic synthesis	Prebiotic synthesis
References	[23]	[40]	[24]	[29]	[25, 41, 42]

Figure 2. Types of membrane forming non-biological amphiphiles. R= alkyl chains, between 7 and 17 carbons and x = repeats between 0 and 3 times.

When examined more closely, the organic matter in meteorites showed fatty acids up to 12 carbons in length. Fatty acids (with chain lengths greater than seven carbons) are known to self assemble into bilayer structures when the pH is adjusted close to the apparent pK_a[23, 24]. This is due to the electrostatic interactions between headgroups when approximately half of the carboxylic groups are deprotonated. Because such bilayers are closely related to a modern cell membrane, they have been studied extensively as the most probable prebiotic container.

Among carboxylic acids, mono-unsaturated fatty acid vesicles (oleic acid and shorter chain derivatives) are the most studied due to their stability at room temperature. They also possess an headgroup considered more plausible for prebiotic amphiphiles than phospholipids. However, mono-unsaturated fatty acids are primarily produced by biological means (to the best of our knowledge, there is no known prebiotic synthesis). In addition, unsaturated fatty acids are sensitive to oxidation, which increases as a function of temperature. Fully saturated fatty acids with long hydrocarbon chains would be more prebiotically plausible, but they are more difficult to work with because they tend to precipitate at low temperatures (20 to 40°C).

Table 1. Fatty acid interactions with the environment

Environmental factor	Positive effects	Negative effects
Large water volumes	None suggested	Dilution of protocell components
Salts	Catalyst for chemical reactions	Destabilization of SCA vesicles
pH	Fatty acid assembly at pKa Non-enzymatic polymerization of Nucleic acids	Destabilization of SCAs at incorrect pH Degradation of biological molecules
High temperature	Nucleic acid strand melting Higher energy for chemical reactions Mixing of protocell	Degradation of temperature sensitive compounds Vesicle instability

The interaction of membranous structures with metabolic networks and information molecules is of primary interest. However, the environmental conditions have been shown to be highly important for vesicle formation specifically using single chain amphiphiles (SCAs). This has imposed an additional challenge on SCA container generation due to stability (or lack there of) under specific conditions, and will be given appropriate attention.

Interaction with External Media

For primordial containers to be functional, the structure itself would have to be relatively inert to environmental solutes. Unlike phospholipids, SCAs tend to be sensitive to a wide range of molecules and conditions. By adjusting environmental parameters, such as pH, salt concentration, and temperature, it is possible to optimize SCA vesicle formation; the pros and cons of each condition are summarized in Table 1. In fact, it may be possible to find a system of mixed amphiphiles that can form bilayers under most pH-, temperature-, or salt concentration-ranges.

Water volume effects

The concentration of an amphiphile that is required to form bilayers (critical vesicle concentration, CVC) depends on hydrocarbon chain length as well as headgroup properties. On early Earth the ability of simple amphiphiles to form bilayers would be highly dependent on the availability of these molecules. If the molecules were originating in a large salty ocean it is unlikely that they would have an opportunity to form into stable structures. It has been shown that the CVC of a pure fatty acid can be decreased by adding a similar length fatty acid derivative such as glycerol mono acyl amphiphile (GMA) or fatty alcohol[24, 37]. Depending on the chain length, the CVC decreases from 4 to 9 times. This indicates that instead of needing 10 mM decanoic acid, upon the addition of a ten carbon GMA, the CVC decreases to 1.2 mM. Similar effects are seen with 8, 12 and 18 carbon fatty acids. Decanol will also decrease the CVC of decanoic acid to a comparable amount. This decrease in the CVC using mixed vesicle compositions could allow vesicles to form at a lower concentration in the early oceans.

Another possibility for overcoming a large dilution is suggested in cycles of dehydration that could have occurred at the land/water interface[38]. Shoreline areas would have allowed for accumulation of organic molecules as areas cycled through wet and dry periods. These molecules could have been concentrated to sufficient numbers to form vesicles during a drought. If these vesicles remained trapped in pools or local "dead zones" of current, they could have been in an ideal location to interact with other prebiotically relevant molecules.

pH sensitivity

Depending on composition, SCA vesicles show a wide range of stability, which is the result of the interaction with the aqueous solution that they are exposed to. As mentioned briefly above, fatty acids are sensitive to the pH of a solution. At low pH the acid forms an oily phase in solution; at high pH the fully deprotonated acids form micellar structures. Vesicles will be formed only when the environmental pH is adjusted as to obtain the correct ratio of protonated to deprotonated fatty acid.

Altering the composition of the surfactants used can lessen pH dependence. By adding a SCA that contains an alcohol headgroup, such as fatty alcohol[24] or GMA[39], high pH stability of vesicles can be observed. This effect is due to non-ionizable hydroxyl group interaction with the deprotonated fatty acids altering the hydrogen bonding or possibly changing the phase transition temperature.

As recently reported, it is possible to stabilize fatty acid bilayers at low pH by adding an amino-headgroup[40]. A 1:1 ratio of decanoic acid/decylamine could form bilayers at pH less than 3 and above 10. Although no stable bilayers were observed at neutral pH, the ability to form structures at extreme pH demonstrates the versatility of these amphiphiles under a variety of conditions.

Polyprenyl compounds also show pH dependent vesicle formation[25, 41, 42]. A wide range of derivatives were tested with single- and multi-branched tails and the formation of vesicles was possible over a range of pH varying from 3 and 13 depending on the exact polyprenyl structure. Polyprenyl phosphates can form vesicles alone, however greater pH stability is obtained by mixing up to 50 mol% of polyprenyl alcohol into the vesicle preparation.

Salt sensitivity

Although the salinity of the early oceans remains unknown, one can assume that salts were present in the "prebiotic soup" with estimates as high as twice the present levels[16]. The presence of salts would have been required for a metabolism centered around metal-ions or to fulfill the RNA world hypothesis. Monovalent salts are known to promote RNA folding and divalent cations, e.g., Mg^{2+}, act as co-factors in ribozyme activity. However, salts are a destabilizing factor for some SCAs [39] even at millimolar concentrations because much of their stability relies on the electrostatic interactions of the headgroups, unlike *two* carbon-chain lipids, which have stronger hydrophobic associations. This observed sensitivity of pure SCA bilayers to salts can be reduced by varying the headgroup composition to such an extent that it becomes possible to maintain bilayer structures at a concentration of salts compatible with RNA ribozyme activity.

For example, vesicles formed from pure decanoic acid will precipitate when cations are added because the bilayer formation and stability rely heavily on hydrogen bonds between protonated and deprotonated headgroups, which is disturbed by cations. The first effective stabilizing amphiphile found was mono-acyl glycerol[39]. Even at concentrations of sea salts comparable to modern oceans, bilayers were stable at a 1:2 ratio of glycerol mono-decanoate/decanoic acid (DA, saturated 10C fatty acid). The glycerol headgroup can likely shield the negative charges on the vesicle surface from cations.

Further salt stabilizations were seen using other amphiphiles. Decanol, when acting as a co-surfactant, decreases the monovalent salt sensitivity even at very high DA to decanol ratios (9:1). Divalent cations, however, still cause the precipitation of such a system. Using a decylamine/decanol system is more successful at creating salt-tolerant vesicles at low pH [40]. At low pH, decylamine is positively charged, which can then repel the cations. At high pH the vesicles become negatively charged and therefore, lose their salt tolerance.

Other, still unknown, surfactants were likely present on early Earth, which could have further stabilized SCA bilayers. Additional research is needed to determine how this salt interaction affects other vesicle properties such as CVC and encapsulation.

High temperature vesicle formation

The effect of temperature is in some ways unique as to affecting primitive chemical systems. The Earth was slowly cooling from above boiling to approximately 50 °C in the Achaean Era when the first evidence of biological compounds appear (~3.5Ga)[16]. Presumably the first life appeared before this time, but the life properties (e.g., lack of skeletal support) and tectonic activity may well have contributed to the absence of any fossil record or its obliteration. We know that the lipid requirements change greatly for an organism to survive at temperatures exceeding 50 °C. Thermophiles today use branched chain hydrocarbons and membrane spanning lipids[43] to increase the stability of their membranes at these higher temperatures. The question of how first life overcame the temperature issue remains open.

Single unsaturated hydrocarbon-chain amphiphiles, such as oleic acid that are stable at elevated temperatures, would be the subject of oxidation by any high-energy radical formed under such conditions. This is one of the reasons why saturated fatty acids are considered more prebiotic. Saturated fatty acids with a hydrocarbon chain length longer than 11C form stable vesicles when the temperature is increased[37, 44]. Thus, pre-Archean vesicles composed of these FAs may have performed much like the unsaturated counterparts, but at elevated temperature.

The effect of long-term temperature exposure was investigated for some saturated short-chain fatty acids as well as oleic acid. As expected, oleic acid broke down after 24h at temperatures >65°C. Similarly, DA vesicles broke down after 24h at only 45°C (in buffered solution containing 100mM phosphate). However, lauric acid and oleic acid remained stable at 45°C, which is a conservative temperature in the context of the emergence of life.

Higher temperatures would allow for reactions requiring higher energy to occur (but would also degrade more fragile molecules). They would also increase bilayer permeability or/and vesicle-vesicle interactions. As a simple SCA mixture can produce more stable structures than a single amphiphile system at room temperature, a different mixture could perform similarly at higher temperatures. The lack of reports here is clearly stemming from the experimental limitations associated with these vesicles that are stable at higher temperatures (like lauric acid) but tend to precipitate at lower temperatures. Performing experiments in which the solutions stay constantly at elevated temperatures is very difficult, although possibly preferable when modeling the origin of life.

Temperature-related fusion

Vesicle-vesicle interaction is interesting from a genetic point of view because while individuality of modern cells is the key to evolution on a relatively long time frame, in simple systems this quality may not have been as essential. The interchange of internal contents as well as bilayer components could have led to greater stability of primitive cells and increase of complexity leading to the emergence of new functions. Indeed, the bilayer could be made more stable by exchanging compositions thereby increasing non-covalent interactions. Meanwhile, the internal contents could be shared so that metabolic catalysts could interact with information carriers that were developed in separate containers, and then mixed without being exposed to the external media.

One of the simplest bilayer forming systems, decanoic acid, does not form vesicles that are stable as individual entities. It was recently shown[37] that at room temperature the interchange of membrane materials between vesicles occurred significantly over 24h and that this interchange was increased with higher temperatures. It was proposed that this amphiphile exchange was due to the fusion of vesicles in the system and that no individuality could reliably be maintained.

Table 2. Solute encapsulation inside of fatty acid vesicles and their derivatives

Molecules	Chain length (Carbons: unsaturated bonds)	Additional stabilizers	References
H^+	18:1	Arginine	[47]
CF	18:1	Long chain normal alcohol, mono-acyl glycerol	[49]
Calcein	10:0, 14:1	mono-acyl glycerol, Long chain normal alcohol	[49]
Pyranine	10:0	Pyrene and acylamine,	[40]
tRNA	10:0, 12:0, 14:1, 18:1	mono-acyl glycerol	[37]
Nucleotides	18:1	None	[51]
	14:1, 18:1	mono-acyl glycerol	[52]
Proteins	14:1, 18:1	mono-acyl glycerol	[51]
short DNA (≤10-mer)	10:0, 14:1, 16:1, 18:1, 18:2	mono-acyl glycerol, Long chain normal alcohol	[52]
long DNA (>10-mer)	10:0, 14:1, 18:1	Long chain normal alcohol and mono-caprin, mono-acyl glycerol	[52]

Increasing the complexity of the membrane can induce greater stability, with less exchange of materials. This can be accomplished by increasing the chain length of the fatty acids or by changing some of the headgroups to allow for increased electrostatic interactions within the bilayer. Increasing the chain length from 10 to 12 carbons decreased the mixing by 10-30%. GMAs were shown to decrease the mixing between two populations up to 60% even at increased temperatures[37]. The implications of this finding are that fatty acids alone may not form stable enough vesicles for the individuality required for an evolving entity and that at higher temperatures this becomes more pronounced.

Permeability of Solutes

The interaction of primitive containers with their environments must include their ability to encapsulate external sources of catalysis, nutrition, and information bearing molecules. Much like modern cells exchange molecules with external media, early compartments should have been able to do the same to allow for the development of metabolism and the genetic code.

Small molecule permeability

The permeability of building blocks into a stable container would be essential for the development of more complex macromolecules (such as proteins or DNA). Permeability has been measured for nucleotides, sugars, amino acids, ions, and dyes through a variety of liposome membranes (Table 2).

These molecules are transported either by a diffusion mechanism (relying on active transport using the flip-flop of amphiphiles in some cases) or by formation of transient pores (due to membrane packing instabilities) [45, 46]. Small molecule permeation is dependent on the amphiphile headgroup and hydrocarbon tail length, as well as on the solute size and charge. Long hydrocarbon chain SCA, such as oleic acid, are capable of containing small charged solutes whereas their short chain counterparts, such as octanoic acid, are not stable enough to hold any molecule.

The smallest ions relevant to modern cell function are hydronium ions, or protons as the ability of maintaining a pH gradient across a membrane can be used to transiently store energy. Bilayer membranes solely formed by fatty acids cannot maintain on their own such a gradient because of the fast flip-flop of fatty acids across a membrane that facilitates H^+ transport. However, it was discovered that by adding arginine externally to oleic acid vesicles this transport could be reduced as the relative impermeability of the membranes to the arginine cation allowed for the accumulation of protons in the internal aqueous volume, thereby creating a proton gradient[47].

It is worth mentioning that other mono-ions, such as alkali metals, [47] have been tested for permeability through oleic acid membranes, and the expected trend of decreased permeability with increasing ionic radius was observed. Using this as a model, it is likely that other ions follow similar trends although headgroup-solute interactions could be detrimental to the stability as discussed above.

Sugars are used as an energy source for modern cells as well as a backbone for genetic polymers and therefore are very important to the understanding of the origins of life. In addition, their permeability allows us to investigate effects of uncharged polar molecule sizes on their permeation through a vesicle bilayer. One of the first studies of fatty acid vesicles proved that glucose that was encapsulated in oleic acid vesicles was protected from an external Glucose-6-phosphate dehydrogenase[48].

More recently, a detailed study of sugar permeability in various chain length fatty acids was published[49]. Sugars of 3 to 6 carbons were measured for permeability in unsaturated fatty acids. All of the sugars followed the general trend of smaller molecules permeating faster, except for one, Ribose. This sugar permeates an order of magnitude faster than any other similarly sized sugar. Computer simulations [50] have demonstrated that interactions between the exocyclic hydroxyl moieties, which are more extensive in ribose, cause a faster permeation through the hydrophobic bilayer. This selective permeability would allow it to accumulate and be protected inside of membranes and also be more readily available for synthetic reactions, perhaps even leading to its incorporation into information polymers.

As building blocks of genetic material, nucleotides have also been measured for permeability in a variety of membrane systems [51, 52]. These studies clearly established the dependence of permeation rates on the solute charges. Whereas natural nucleotides are relatively slowly permeating into single-chain amphiphile vesicles, activated nucleotides, which are zwitterionic, do it significantly more rapidly due to the reduction in charge on the molecule.

The same dependence on charge can be observed for amino acids[53, 54]. Because of their zwitterionic nature, they are usually charged molecules. However, amino acids can become neutral and then slowly pass the hydrophobic barrier. This charge dependent permeability could have been advantageous by leading to accumulation of the charged amino acids within the internal volume of protocell compartment.

The stabilization of the hydrophobic barrier by increasing their compositional complexity will result in an alteration of their permeability toward solutes, whose extent depends on the hydrocarbon chain length but also on the electrostatic interactions between the headgroups of the amphiphiles. Decanoic acid vesicles cannot entrap dyes or tRNA[37] but the addition of decanol will allow the clean separation of both (Figure 3), while the addition of GMA will allow tRNA encapsulation but not dye. Thus, it seems possible to modify the bilayer characteristics and achieve a new "molecular weight cut off" by simply modifying the SCA composition.

Large molecule permeability: nucleic acid polymer retention

Nucleic acid polymers play an essential role in the function of living cells today, and it seems probable that, as proposed in the RNA world hypothesis, catalytic and informational RNA polymers played essential roles in the emergence of the first living entities. Thus, the study of the interaction of these polymers with primitive containers is essential. The encapsulation of nucleic acid polymer would be beneficial on several levels: to co-locate several catalysts creating catalytic networks, to protect these polymers from degradative agents, and finally to facilitate their replication.

Oleic acid vesicles, among others, are capable of encapsulating nucleic acid strands with long retention time (>24h). It was also recently shown that the rate of leakage from the membrane was dependent on the size of the DNA incorporated with smaller oligomers leaking faster than larger ones[52], and on the stability amphiphilic bilayer, with oleic acid being almost impermeable and decanoic acid with decanol and glycerol monodecanoate allowing faster leakage.

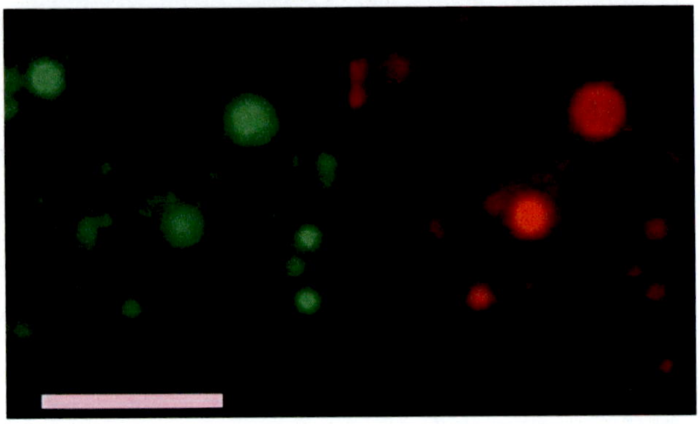

Figure 3. Pictures of vesicles. Green pyranine dye is trapped inside of 2:1 DA/DOH vesicles, while Nile red stains the membrane only. Scale bar is 25μm

One exciting study in the quest for building a protocell has established that template directed polymerization from activated nucleotides can be achieved non-enzymatically inside SCA vesicles composed of fatty acid with hydrocarbon chains as short as 10 carbons[52]. Activated nucleotides that were added to the external volume of a vesicle solution containing a poly(C) template had to permeate across the bilayer to be incorporated into a poly(G) strand.

This discovery suggests that the replication of template inside of a closed container is possible and thus provided a population of vesicles with various information ("genetic") content, therefore a selection process might have occurred. However, this system has a few limitations. First, the polymerization was of a homopolymer in this population and only the cognate nucleotide was added. Thus, it contained no real genetic information. The prebiotic availability of activated nucleobases on the early Earth is also disputed. However, the use of 10C SCA as an encapsulating system for template-directed polymerization is a step toward creating a working protocell.

Temperature related permeability

The permeability has indeed been shown to increase in FA vesicles with increasing temperature. Oleic acid showed leakage of a small fluorescent dye (carboxyfluorescein) at temperatures of greater than 30°C at times greater than 24h[37]. The leakage of the same molecule from a 12 carbon mixture was much faster and at 45°C leaked to equilibration after only 1h. Notably, the oleic acid can be stabilized with the addition of an 18 carbon GMA and cause the temperature induced leakage to be reduced, but not entirely prevented.

Encapsulation, as discussed above, is vital for the maintenance of a chemical metabolism. Mansy and Szostak, recently discovered that it was possible to maintain encapsulated oligomers (3, 4, and 10 polyA) for short periods of time (<10h) at temperatures above 0°C but less than 100°C[55]. At these temperatures, the observed leakage was correlated with the length of the "information" molecules and decreased upon the addition of GMAs.

Thus, variation of temperature would allow a protocell to cycle through stages of high to low exchange with the environment, as well as power its metabolism [55]. Short periods (one minute) of exposure to 90°C and then cooling to 20°C are sufficient to successfully dehydribize encapsulated double stranded 10-mer DNA. Thus, such temperature fluctuation could have promoted the replication of short nucleic acids without the use of the complex protein structures needed in modern cells which serve to "unzip" the helical strands. Using this process along with polymerization it would have been possible to create multiple generations of information molecules inside of a single container.

Metabolism Incorporation

For the early cell to be efficient, it must have possessed semi-permeable compartment boundaries that allow transport of nutrients into the internal content where its metabolic network utilized or metabolized them into its building blocks (resulting in its maintenance or growth) while waste products would diffuse back out of the protocell. Catalytic systems, such as ribozymes, single protein enzymes, and even entire enzyme systems have been encapsulated inside of SCA vesicles to demonstrate that cellular tasks can be performed in these containers. In most investigations, the use of enzymes is clearly considered as palliative for the lack of more simple catalytic polymers, such as those proposed by Szostak *et al.* [2].

The RNA world hypothesis spurred interest in catalytic nucleic acid material. In SCA studies, a hammerhead ribozyme (62 nucleobases long), which is known to self-cleave in the presence of Mg^{2+} (4 mM), was studied inside of glycerol mono-myristoleate/myristoleic acid vesicles [56]. The addition of Mg^{2+} in the external medium induced cleavage of the encapsulated ribozyme within the compartment. That is, the Mg^{2+} had to permeate across the bilayers and reach internal concentration compatible with the ribozyme activity. GMA imparted a salt stability absent in pure FA vesicles, which was paramount for the ribozyme activity.

Other metabolic reactions have demonstrated that an enzyme inside of a membrane is effectively protected from degradative agents. Catalase was shown to be successfully encapsulated and protected from protease, which was added to the external vesicle solution. The addition of H_2O_2 to the external volume and subsequent diffusion into the vesicles allowed for the enzymatic conversion into water and O_2. The barrier function of this experiment exemplifies the semi-permeable nature of single chain amphiphiles[24].

As discussed earlier, the non-enzymatic template-directed polymerization of poly(G) was achieved in SCA vesicles without truly leading to information generation in terms of sequences and nucleic acid length. However, it is also possible to create a metabolic network, which (re)produces long nucleic acid fragments. Polynucleotide phosphorylase (PNPase) [51], which is capable of forming long poly(A) strands from adenosine diphosphate (ADP), was shown to function inside of oleic acid vesicles using ADP molecules that were added externally, thus had to permeate the membrane for polymerization to occur. Furthermore, a template-directed RNA production, i.e., information replication, inside of SCA vesicles has also been achieved using the enzyme Q-ß replicase[57]. In the preparation mixture, nucleotides and the enzyme along with an RNA template were encapsulated together in oleic acid vesicles. The replication of the template was again observed even in the presence of degradative agents (EDTA and Ribonuclease A) in the external medium. Thus, a functional metabolic system inside of a SCA container can be protected from degradation and is capable of generating potential information.

However, for a truly functional protocell, the components must rely on each other for replication. In the cases cited above, the host compartment is not linked to either the metabolic assembly or the information polymer product. In fact there is no real benefit to any of the components due to the presence of any other part. Furthermore, the absence of proteins on the early Earth has been postulated long ago, thus the PNPase or Q-ß replicase systems should be considered laboratory models. Nonetheless, the conversion of a nutrient to a more complex polymer is a proof of principle that encapsulated metabolic material can produce nucleic acid polymers inside of SCA containers.

The last main purpose of a metabolism is to harvest energy from the environment and convert it into useful molecules for the cell. While this is not a simple process, one of the known primitive molecules may have been efficient at achieving this end. PAHs were found in meteorite samples as well as in modeled prebiotic synthesis[12]. These molecules can be made to integrate into amphiphile membranes and then are capable of harvesting light and creating a proton gradient. If this gradient could be coupled to high-energy molecule generation, a protocell metabolism could be developed around this class of molecule[7].

Linking Metabolism, Information and Containment to Achieve Protocell Life Cycles

The presence of metabolic, informational, and containment components need to form an inter-dependent relationship that is crucial to create a truly functional protocell. Some more recent work has been focused on metabolically generated containers that also include a information ("genetic") component. The presence of all the protocell components is necessary to impart self-replicating properties on these chemical systems. These new more complex protocell models will allow researchers to study life cycles of a protocell, and related processes such as amphiphile generation, insertion into a bilayer, and division.

One approach to achieve an information to metabolic to containment linkage is exemplified by the minimal cell concept[58]. This method uses existing genes to produce proteins, which will produce bilayer-forming molecules. Although the highly evolved set of genes employed in this type of experiment is clearly not prebiotic, the use of an existing genetic/metabolic network simplifies the process of creating the living from the non-living. Using phospholipid membranes, the Luisi group has encapsulated a functional cell-free genetic/metabolic system, which can produce proteins involved in the synthesis of phosphatic acid[59]. Although these synthesized proteins were not yet capable of producing membrane-forming material at a rate sufficient to measure the container growth, this experiment established the feasibility of internal amphiphile production and thus represents a significant step in linking the three components of minimal life.

Our group has taken a slightly different approach, which uses a metal-centered lipophilic catalyst to photochemically produce bilayer-forming FAs [60] from a precursor species that is not bilayer-forming on its own. The information ("genetic") component is directly involved in the metabolic reaction as a reducing agent (regenerated multiple times during reaction studies). At the time, a single modified nucleobase (obviously non-replicable) was used as a model for a true information polymer.

The physical link between container, the bilayers themselves, and information/metabolism can be modulated through hydrophobic anchors (hydrocarbon chains) on the information molecules and the metabolic complexes (unpublished observations). Through these multiple interactions, it is possible to get a metabolism-information-container complex, which will react with precursor molecules to form more bilayer building blocks. In due course, nucleic acid oligomers that can be replicated by the same metabolism will replace the modified nucleobase. To date, no reproduction of the metal complex catalyst is possible.

Growth and division

The final step in creating a successful self-reproducing chemical system is to replicate the system as a whole and evolve it. Growth of a minimal cell would be triggered by the conversion of nutrients into minimal cell building blocks, which would be incorporated into the cell; division could then occur to split the initial protocell into daughter cells. These two steps are essential for selection and therefore evolution.

The growth process for a protocell can occur through a few pathways[61]. Vesicles are capable of incorporating newly formed amphiphiles in the form of monomers into the bilayer. They are also capable of scavenging micelles from the solution and transforming them into bilayers. The third possible mechanism for vesicle growth is through fusion events with other vesicles[62].

Experimentally growth of membranes has been shown by the slow injection of micelles, which incorporate into the bilayers of preformed vesicles and increase vesicle size. This process has been demonstrated with vesicles composed of oleic acid and myristoleic acid, as well as decanoic acid. The impact of the incorporation on the vesicle size in a population depends on the speed of addition[63]. Slow addition of small amounts of micelles leads to a more even distribution of the new material into the pre-existing vesicles. Fast addition can cause the formation of new vesicles from the material or the incorporation of large amounts of material into fewer vesicles.

Osmotic stress can also cause growth by vesicle-vesicle interaction. In 2004, Chen, *et al*[62], showed that putting osmotic stress on a fatty acid vesicle would cause it to fuse with isotonic vesicles to reduce the pressure. This suggests that a highly functional protocell that would rapidly produce entrapped metabolic material would be able to scavenge membrane-forming material from proximal less active vesicles to adjust its size.

This division process is still highly speculative. Such a process should occur spontaneously once the initial vesicles have grown to a certain size, and result in the formation of smaller structures having the same properties as the initial systems. Several mechanisms for division have been suggested. Mechanical division could be obtained by submitting the grown vesicles to an external force [64]. Intrinsic division forces are also possible much like today's cell division processes[65].

Extrusion processes [66] can be categorized as a classic laboratory example of mechanical division. This method uses high pressure to force vesicles through very small pores causing a reduction in size. On the early Earth, this process could have been achieved by a moving body of water containing a protocell being pushed through a porous rock, although its likelihood has been questioned. Recently, division of SCA vesicles was obtained by gentle environmental stress[64], a process that might be plausible. First, oleic acid vesicles were "fed" with oleic acid micelles inducing their growth, division was then induced by shaking them gently or applying airflow. This mechanical division method did not cause the leakage of internal components, and the cycle of growth and division was repeated several times. Forces similar to this were likely on the turbulent early Earth considering the estimated volcanism and meteor impacts.

The other form of division that is possible is due to instability in the membrane leading to division by either budding off small daughter cells or a symmetric split of the vesicle. There is some evidence of asymmetric phase separation of two lipids in a vesicle followed by the budding off of a daughter cell[65, 67]. It has also been shown that changing the local environment rapidly will cause the budding of a phospholipid bilayer and eventually division.

Growth and division are essential for the natural selection process to occur, which is one of the fundamental qualities of life. The growth that has been shown to date is caused by a scavenge process from the external media, as opposed to a metabolic production. Also for a protocell, random division events like shaking are imperfect. For a functional protocell these processes need to be under the control of a genetic and/or metabolic system.

CONCLUSIONS

The formation of stable containers under various conditions is possible using a range of materials, mainly single chain amphiphiles. The behavior and stability under a wide range of conditions demonstrates the versatility of single-chain amphiphile structures. From the studies presented in this review one can surmise that the large collection of single chain amphiphile molecules, likely present on the early Earth, could have contributed to the formation of protocell containers with the necessary properties to withstand plausible environmental stresses.

Aside from a container, a form of inheritable information and a metabolic (catalytic and energy harvesting) subsystem, as well as their inter-connection were required to start life. To date, it has been shown that separately all parts (taking various forms) could be primordial and so current research is now focused on connecting these parts into a whole protocell. The so far established positive interactions of "primitive" membrane models with metabolic and information ("genetic") molecules represent promising steps towards creating a self-reproducing chemical system. However, an interlinked functional system has yet to be reported. The successful construction of such a system will offer a platform to truly test one essential feature of life: its evolvability.

OUTLOOK

The possibility of using higher temperatures to model the origin of life needs to be investigated in depth, as elevated temperatures seem plausible on the early Earth. Their effects on protocell components may have been quite large in terms of system composition and processes, as well as lifetimes. Studies of protocell models under elevated temperatures may reveal interesting, but overlooked properties of early chemical systems. Current research has been focused on container related temperature effects but it would be interesting to model genetic and metabolic systems at these temperatures as well.

One of the major open issues in the search for a protocell resides in the metabolic processes and their components whereas potential candidates for information and container generation can be proposed. RNA catalysts and simple metal ions systems have been suggested as a metabolic candidate. Obviously, the nature of the metabolic molecules will have an impact on their replicating behavior, thus on protocell self-replication. If "metabolic" catalysis was predominantly carried out by metal ions, it is then conceivable that no metabolic reproduction was necessary. However, the existence of a metabolic RNA-centered system and its subsequent replication represents an interesting alternative to metal-ions, but has to be demonstrated yet.

Finally, another aspect to be emphasized is non-covalent interactions. These weak bonds are just as important for the cellular supermolecular assemblies as the strong covalent bonds. From the effects of salts on membranes or the hydrophobic interaction of an energy-harvesting molecule with a vesicle surface, one can surmise that these weak interactions would have been as important for primitive systems as Watson and Crick base pairing.

ACKNOWLEDGMENTS

This work was supported by the Center for Fundamental Living Technology (FLinT) at the University of Southern Denmark. We would also like to thank David W. Deamer for his comments and discussion of the material throughout the preparation of this manuscript.

LIST OF ABBREVIATIONS

ADP	adenosine diphosphate
CVC	critical vesicle concentration
DA	decanoic acid
FA	fatty acid
GMA	glycerol mono-acyl amphiphile
PAH	polycyclic aromatic hydrocarbon
PNPase	polynucleotide phosphorylase
poly(N)	nucleic acid homopolymers (N=A, C, G, U, or T)
SCA	single chain amphiphile;
tRNA	transfer RNA;

REFERENCES

[1] Gilbert, W. (1986). Origin of Life - the Rna World. *Nature*, *319*, 618-618.

[2] Szostak, J. W., Bartel, D. P. & Luisi, P.L. (2001). Synthesizing life. *Nature*, *409*, 387-390.

[3] Fox, S. W., Bahn, P. R., Dose, K., Harada, K., Hsu, L., Ishima, Y., Jungck, J., Kendrick, J., Krampitz, G., Lacey, J. C., Matsuno, K., Melius, P., Middlebrook, M., Nakashima, T., Pappelis, A., Pol, A., Rohlfing, D. L., Vegotsky, A., Waehneldt, T. V.,

Wax, H. & Yu, B. (1994). Experimental Retracement of the Origins of a Protocell - It Was Also a Protoneuron. *J Biol Phys*, *20*, 17-36.

[4] Watson, E. B. & Harrison, T. M. (2005). Zircon thermometer reveals minimum melting conditions on earliest Earth. *Science*, *308*, 841-844.

[5] Segre, D., Ben-Eli, D., Deamer, D. & Lancet, D. (2001). The Lipid World. *Origins of Life and Evolution of the Biosphere*, *31*, 27.

[6] Joyce, G. F. (2002). The antiquity of RNA-based evolution. *Nature*, *418*, 214-221.

[7] Ehrenfreund, P., Rasmussen, S., Cleaves, J. & Chen, L. (2006). Experimentally tracing the key steps in the origin of life: The aromatic world. *Astrobiology*, *6*, 490-520.

[8] Shapiro, R. (2006). Small molecule interactions were central to the origin of life. *Q Rev Biol*, *81*, 105-125.

[9] Weber, A. L. (1996). The Metabolic World: Sugars as an Energized Carbon Substrate for Prebiotic and Biotic Synthesis. *Origins of Life and Evolution of the Biosphere*, *26*, 2.

[10] Miller, S. L. & Urey, H. C. (1959). Organic Compound Synthesis on the Primitive Earth. *Science*, *130*, 7.

[11] McCollom, T. M., Ritter, G. & Simoneit, B. R. (1999). Lipid synthesis under hydrothermal conditions by Fischer-Tropsch-type reactions. *Origins of Life and Evolution of the Biosphere*, *29*, 14.

[12] McCollom, T. M. & Simoneit, B. R. T. (1999). Abiotic formation of hydrocarbons and oxygenated compounds during thermal decomposition of iron oxalate. *Origins of Life and Evolution of the Biosphere*, *29*, 167-186.

[13] Chyba, C. & Sagan, C. (1992). Endogenous Production, Exogenous Delivery and Impact-Shock Synthesis of Organic-Molecules - an *Inventory for the Origins of Life. Nature*, *355*, 125-132.

[14] Chyba, C. F., Thomas, P. J., Brookshaw, L. & Sagan, C. (1990). Cometary Delivery of Organic-Molecules to the Early Earth. *Science*, *249*, 366-373.

[15] Simoneit, B. R. T., Rushdi, A. I. & Deamer, D. W. (2007). Abiotic formation of acylglycerols under simulated hydrothermal conditions and self-assembly properties of such lipid products. *Advances in Space Research*, *40*, 1649-1656.

[16] Knauth, L. P. (2005). Temperature and salinity history of Precambrian ocean: implications for the course of microbial evolution. Palaeogeography, Palaeoclimatology, *Palaeoecology*, *219*, 53-69.

[17] Ferris, J. P., Joshi, P., Aldersley, M., Miyakawa, S., Huang, W. H. & Wang, K. J. (2006). Mineral catalysis and prebiotic synthesis. *Origins of Life and Evolution of the Biosphere*, *36*, 212-212.

[18] Liu, R. & Orgel, L. E. (1998). Polymerization on the rocks: beta-amino acids and arginine. *Origins of Life and Evolution of the Biosphere*, *28*, 245-257.

[19] Hill, A. R., Bohler, C. & Orgel, L. E. (1998). Polymerization on the rocks: Negatively-charged alpha-amino acids. *Origins of Life and Evolution of the Biosphere*, *28*, 235-243.

[20] Orgel, L. E. (1998). Polymerization on the rocks: Theoretical introduction. *Origins of Life and Evolution of the Biosphere*, *28*, 227-234.

[21] Hanczyc, M. M., Fujikawa, S. M. & Szostak, J. W. (2003). Experimental models of primitive cellular compartments: Encapsulation, growth, and division. *Science*, *302*, 618-622.

[22] Dobson, C. M., Ellison, G. B., Tuck, A. F. & Vaida, V. (2000). Atmospheric aerosols as prebiotic chemical reactors. *P Natl Acad Sci*, USA, *97*, 5.

[23] Hargreaves, W. R. & Deamer, D. W. (1978). Liposomes from Ionic, Single-Chain Amphiphiles. *Biochemistry*, *17*, 3759-3768.

[24] Apel, C. L., Deamer, D. W. & Mautner, M. N. (2002). Self-assembled vesicles of monocarboxylic acids and alcohols: conditions for stability and for the encapsulation of biopolymers. *Biochimica Et Biophysica Acta-Biomembranes*, *1559*, 1-9.

[25] Takajo, S., Nagano, H., Dannenmuller, O., Ghosh, S., Albrecht, A.M., Nakatani, Y. & Ourisson, G. (2001). Membrane properties of sodium 2-and 6-(poly)prenyl-substituted polyprenyl phosphates. *New J Chem*, *25*, 917-929.

[26] Hargreaves, W. R., Mulvihill, S. J. & Deamer, D. W. (1977). Synthesis of Phospholipids and Membranes in Prebiotic Conditions. *Nature*, *266*, 78-80.

[27] Rushdi, A. I. & Simoneit, B. R. T. (2001). Lipid formation by aqueous Fischer-Tropsch-type synthesis over a temperature range of 100 to 400 degrees C. *Origins of Life and Evolution of the Biosphere*, *31*, 103-118.

[28] Ahmed, I. R. & Simoneit, B. R. (2000). Lipid formation by aqueous fischer-tropsch-type synthesis over a temperature range of 100-400 °C. *Origins of Life and Evolution of the Biosphere*, *31*, 16.

[29] Apel, C. L. & Deamer, D. W. (2005). The formation of glycerol monodecanoate by a dehydration/condensation reaction: Increasing the chemical complexity of amphiphiles on the early earth. *Origins of Life and Evolution of the Biosphere*, *35*, 323-332.

[30] Yamagata, Y., Watanabe, H., Saitoh, M., & Namba, T. (1991). Volcanic production of polyphosphates and its relevance to prebiotic evolution. *Nature, 352*, 516-519.

[31] Chyba, C. F. (1993). The Violent Environment of the Origin of Life - Progress and Uncertainties. *Geochimica Et Cosmochimica Acta*, *57*, 3351-3358.

[32] Pizzarello, S., Huang, Y. & Alexandre, M. R. (2008). Molecular asymmetry in extraterrestrial chemistry: Insights from a pristine meteorite. *P Natl Acad Sci*, USA, *105*, 3700-3704.

[33] Pizzarello, S. & Holmes, W. (2009). Nitrogen-containing compounds in two CR2 meteorites: N-15 composition, molecular distribution and precursor molecules. Geochimica Et *Cosmochimica Acta*, *73*, 2150-2162.

[34] Deamer, D. W. & Pashley, R.M . (1989). Amphiphilic Components of the Murchison Carbonaceous Chondrite - Surface-Properties and Membrane Formation. *Origins of Life and Evolution of the Biosphere*, *19*, 21-38.

[35] Dworkin, L. P., Deamer, D. W., Sandford, S. A. & Allamandola, L. J. (2001). Self-assembling amphiphilic molecules: Synthesis in simulated interstellar/precometary ices. *P Natl Acad Sci*, USA *98*, 815-819.

[36] Anders, E. (1989). Pre-Biotic Organic-Matter from Comets and Asteroids. *Nature*, *342*, 255-257.

[37] Maurer, S. E., Deamer, D. W., Boncella, J. M. & Monnard, P. A. (2009). Chemical evolution of amphiphiles: glycerol monoacyl derivatives stabilize plausible prebiotic membranes. *Astrobiology 9*, 979-987.

[38] Deamer, D., Dworkin, J. P., Sandford, S. A., Bernstein, M. P. & Allamandola, L. J. (2002). The first cell membranes. *Astrobiology*, *2*, 371-381.

[39] Monnard, P. A., Apel, C. L., Kanavarioti, A. & Deamer, D. W. (2002). Influence of ionic inorganic solutes on self-assembly and polymerization processes related to early forms of life: Implications for a prebiotic aqueous medium. *Astrobiology*, *2*, 139-152.

[40] Namani, T. & Deamer, D.W. (2008). Stability of model membranes in extreme environments. *Origins of Life and Evolution of the Biosphere*, *38*, 329-341.

[41] Streiff, S., Ribeiro, N., Wu, Z., Gumienna-Kontecka, E., Elhabiri, M., Albrecht-Gary, A., Ourisson, G. & Nakatani, Y. (2007). ''Primitive'' Membrane from Polyprenyl Phosphates and Polyprenyl Alcohols. *Chemistry and Biology*, *14*, 313-319.

[42] Gotoh, M., Miki, A., Nagano, H., Ribeiro, N., Elhabiri, M., Gumienna-Kontecka, E., Albrecht-Gary, A., Schmutz, M., Ourisson, G. & Nakatani, Y. (2006). Membrane Properties of Branched Polyprenyl Phosphates, Postulated as Primitive Membrane Constituents. *Chemistry and Biodiversity*, *3*, 434-455.

[43] De Rosa, M., Trincone, A., Nicolaus, B. & Gambacorta, A. (1991). Archeabacteria: lipids, membrane structures, and adaptations to environmental stresses. In Life under extreme conditions : biochemical adaptation, G. Di Prisco and Federation of European *Biochemical Societies. Meeting, eds*., (Berlin ; New York: Springer-Verlag), vii, 144.

[44] Skurtveit, R., Sjoblom, J. & Hoiland, H. (1989). Emulsions under Elevated-Temperature and Pressure Conditions .1. The Model System Water Hexadecanoic Acid Sodium Hexadecanoate Decane at 70-Degrees-C. *J Colloid Interf Sci*, *133*, 395-403.

[45] Paula, S. & Deamer, D. W. (1999). Membrane permeability barriers to ionic and polar solutes. *Membrane Permeability*, *48*, 77-95.

[46] Paula, S., Volkov, A. G., VanHoek, A. N., Haines, T. H. & Deamer, D. W. (1996). Permeation of protons, potassium ions, and small polar molecules through phospholipid bilayers as a function of membrane thickness. *Biophysical Journal*, *70*, 339-348.

[47] Chen, I. A. & Szostak, J. W. (2004). Membrane growth can generate a transmembrane pH gradient in fatty acid vesicles. *P Natl Acad Sci*, USA *101*, 7965-7970.

[48] Gebicki, J. M. & Hicks, M. (1973). Ufasomes Are Stable Particles Surrounded by Unsaturated Fatty-Acid Membranes. *Nature*, *243*, 232-234.

[49] Sacerdote, M. G. & Szostak, J. W. (2005). Semipermeable lipid bilayers exhibit diastereoselectivity favoring ribose. *P Natl Acad Sci*, USA *102*, 6004-6008.

[50] Wei, C. Y. & Pohorille, A. (2009). Permeation of Membranes by Ribose and Its Diastereomers. *J Am Chem Soc*, *131*, 10237-10245.

[51] Walde, P., Goto, A., Monnard, P.A., Wessicken, M. & Luisi, P. L. (1994). Oparins Reactions Revisited - Enzymatic-Synthesis of Poly(Adenylic Acid) in Micelles and Self-Reproducing Vesicles. *J Am Chem Soc*, *116*, 7541-7547.

[52] Mansy, S. S., Schrum, J. P., Krishnamurthy, M., Tobe, S., Treco, D. A. & Szostak, J. W. (2008). Template-directed synthesis of a genetic polymer in a model protocell. *Nature*, *454*, 122-U110.

[53] Chakrabarti, A. C. & Deamer, D. W. (1992). Permeability of Lipid Bilayers to Amino-Acids and Phosphate. *Biochim Biophys Acta*, *1111*, 171-177.

[54] Chakrabarti, A.C . & Deamer, D. W. (1994). Permeation of Membranes by the Neutral Form of Amino-Acids and Peptides - Relevance to the Origin of Peptide Translocation. *Journal of Molecular Evolution*, *39*, 1-5.

[55] Mansy, S. S. & Szostak, J. W. (2008). Thermostability of model protocell membranes. *P Natl Acad Sci*, USA, *105*, 13351-13355.

[56] Chen, I. A., Salehi-Ashtiani, K. & Szostak, J. W. (2005). RNA catalysis in model protocell vesicles. *J Am Chem Soc*, *127*, 13213-13219.
[57] Oberholzer, T., Wick, R., Luisi, P.L. & Biebricher, C. K. (1995). Enzymatic Rna Replication in Self-Reproducing Vesicles - an Approach to a Minimal Cell. *Biochem Bioph Res Co*, *207*, 250-257.
[58] Luisi, P. L., Ferri, F. & Stano, P. (2006). Approaches to semi-synthetic minimal cells: a review. *Naturwissenschaften*, *93*, 1-13.
[59] Murtas, G., Kuruma, Y., Bianchini, P., Diaspro, A. & Luisi, P. L. (2007). Protein synthesis in liposomes with a minimal set of enzymes. *Biochem Bioph Res Co*, *363*, 12-17.
[60] DeClue, M. S., Monnard, P. A., Bailey, J. A., Maurer, S. E., Collis, G. E., Ziock, H. J., Rasmussen, S. & Boncella, J. M. (2009). Nucleobase Mediated, Photocatalytic Vesicle Formation from an Ester Precursor. *J Am Chem Soc*, *131*, 931-933.
[61] Hanczyc, M. M. & Szostak, J. W. (2004). Replicating vesicles as models of primitive cell growth and division. *Current Opinion in Chemical Biology*, *8*, 660-664.
[62] Chen, I. A., Roberts, R. W. & Szostak, J. W. (2004). The emergence of competition between model protocells. *Science*, *305*, 1474-1476.
[63] Chen, I. A. & Szostak, J. W. (2004). A kinetic study of the growth of fatty acid vesicles. *Biophysical Journal*, *87*, 988-998.
[64] Zhu, T. F. & Szostak, J. W. (2009). Coupled Growth and Division of Model Protocell Membranes. *J Am Chem Soc*, *131*, 5705-5713.
[65] Inaoka, Y. & Yamazaki, M. (2007). Vesicle fission of giant unilamellar vesicles of liquid-ordered-phase membranes induced by amphiphiles with a single long hydrocarbon chain. *Langmuir*, *23*, 720-728.
[66] Olson, F., Hunt, C. A., Szoka, F. C., Vail, W. J. & Papahadjopoulos, D. (1979). Preparation of Liposomes of Defined Size Distribution by Extrusion through Polycarbonate Membranes. *Biochim Biophys Acta*, *557*, 9-23.
[67] Baumgart, T., Hess, S. T. & Webb, W. W. (2003). Imaging coexisting fluid domains in biomembrane models coupling curvature and line tension. *Nature*, *425*, 821-824.

In: Astrobiology: Physical Origin, Biological Evolution... ISBN: 978-1-60741-290-8
Editors: S. Hegedus, J. Csonka, pp. 81-95

Chapter 4

MICROBIAL EVOLUTION AND ADAPTATION IN ICY WORLDS

***Alberto Alcazar*[1], *Laura Garcia-Descalzo*[2] *and Cristina Cid*[*2]**

[1] Department of Investigation, Hospital Ramon y Cajal, 28034 Madrid, Spain

[2] Microbial Evolution Laboratory, Center for Astrobiology (CSIC-INTA), 28850 Torrejón de Ardoz, Madrid, Spain

ABSTRACT

There is considerable interest in astrobiology in investigating icy worlds and the microbial forms that thrive in extreme environments, especially under those cold conditions that can provide a model for some extraterrestrial environments. After the successful missions Galileo and Cassini-Huygens to Jupiter and Saturn systems respectively this priority has been focused on the research of environments potentially habitable in some icy worlds such as Europa, Encelado or Titan.

On Earth, the recent discovery of cold-tolerant microorganisms in glaciated and permanently frozen environments has broadened the known range of environmental conditions which support microbial life. These microorganisms, known as psychrophiles are defined as organisms having an optimal temperature for growth at about 15ºC or lower, a maximal temperature for growth at about 20ºC, and a minimal temperature for growth at 0ºC or below. The lowest temperature limit for life seems to be around -20ºC, which is the value reported for bacteria living in permafrost soil and in sea ice. Terrestrial models of extraterrestrial icy worlds are being intensively studied. Among them, bacterial communities from Arctic and Antarctic permafrost, subglacial lakes and high mountains are considered representative of these environments in which psychrophile bacteria are the unique inhabitants. When compared to other known prokaryotes, psychrophilic bacteria posses many unique qualities and molecular mechanisms that allow their adaptation to cold environments. In this regard, some psychrophiles have been found to contain polyunsaturated fatty acids in their plasma membranes, which generally do not occur in other prokaryotes. Further, some of them use enzymes that continue to function at near freezing ambient temperatures. And finally, psychrophilic bacteria are able to

[*] Corresponding author: E-mail: cidsc@inta.es Fax: +34-91-520-10 74

produce proteins that are stable at cold temperatures such as heat shock proteins (Hsps) and cold shock proteins (Csps). This chapter will review the general trends concerning the biodiversity and ecology of psychrophilic microorganisms and their molecular mechanisms of adaptation and evolution, as well as summarize our latest results on this topic in samples from icy environments.

Keywords: cold adaptation, permafrost, extremophiles, psychrophiles, microbial evolution, microbial ecology, biodiversity, molecular mechanisms, polyunsaturated fatty acids, heat shock proteins, cold shock proteins, protein interactions.

1. INTRODUCTION

In the past decades, our concept of a potentially habitable world has broadened from the narrow range between Venus and Mars to cover most of the Solar System and beyond. Two primary factors have contributed to broadening our concept of habitable world. First, there has been an explosion of research on the ability of microorganisms, especially Bacteria and Archaea to grow in extreme environments with respect to temperature, salinity, acidity, radiation and pressure. Environments that would have been considered inhospitable for life a few decades ago have been found thriving with life. The other factor in broadening our concept of a hospitable world is a better understanding of the environments within which liquid water can exist. For example, there is evidence that the moon of Pluto, Charon, may be hot enough to produce liquid water, and today in the planetary sciences, the first step in the search for life is often the search for liquid water (Marion *et al.*, 2003).

The broad interdisciplinary character of astrobiology compels us to strive for the most comprehensive and inclusive understanding of biological, planetary and cosmic phenomena (Des Marais *et al.*, 2003). An interdisciplinary research is required that combines molecular biology, ecology, planetary science, astronomy, information science, space exploration technologies, and related disciplines. The aim of this chapter is to summarize, from a microbiological point of view, what we know about cold adaptation and evolution of psychrophilic microorganisms in order to better understand how icy environments can determine the mechanisms and environmental limits of life.

2. INTEREST IN ASTROBIOLOGY

Astrobiology addresses some basic questions that have been asked in various ways for generations: How did life begin and evolve, does life exist elsewhere in the universe, and what is the future of life on Earth and beyond? Accordingly, astrobiology embraces the search for potentially inhabited planets further than our Solar System, the exploration of Mars and the outer planets, laboratory and field investigations of the origins and early evolution of life, and studies of the potential of life to adapt to future challenges, both on Earth and in space (Des Marais *et al.*, 2003).

There is considerable interest in astrobiology in investigating icy worlds and the microbial forms that thrive in extreme environments, especially under those cold conditions that can provide a model for some extraterrestrial environments. After the successful missions Galileo and Cassini-Huygens to Jupiter and Saturn systems respectively this priority has been focused on the research of environments potentially habitable in some icy worlds such as Europa, Encelado or Titan (Figure 1). For example, Europa is a cold, ice-covered moon of Jupiter that might at first seem inhospitable for life. But there is abundant evidence for the presence of a subsurface briny ocean (Stevenson, 2000). The putative ocean of Europa has focused considerable attention on its possible habitats for life (Chyba, 2000; Chyba & Phillips, 2001).

Also Titan, the sixth and largest moon of the planet Saturn, has shown itself to be an intriguing object for study, with a variety of unusual candidate materials on its surface, such as water-ammonia and other ices, hydrocarbons, and tholins. Titan is very complex geologically, and features found include large craters, cryovolcanoes, mountains, braided and sinuous channels, vast fields of dunes, giant lakes and seas of liquid hydrocarbons. It is thought to be made largely of ice. Some of that ice may melt during meteor impacts or in underground processes, producing "ice volcanoes" that emit a "lava" containing ammonia mixed with water. Observations of Titan by the Cassini spacecraft suggest the presence of cryovolcanism on the moon's surface. Cryovolcanism has important astrobiological implications, as it provides a means of exposing Titan's organics to liquid water, transforming hydrocarbons and nitriles into more evolved and oxidized prebiotic species. Titan appears to be very Earth-like in its geology, despite the very different surface conditions and composition, and has been called "the Earth of the outer solar system". For almost thirty years, scientists have known that complex carbon compounds called tholins exist on comets and in the atmospheres of the outer planets. Theoretically, tholins might interact with water in a process called hydrolysis to produce complex molecules similar to those found on the early Earth. Could tholins formed in Titan's atmosphere react with liquid water temporarily exposed by meteor impacts or ice volcanoes to produce potentially prebiotic complex organic molecules — before the water freezes? Liquid water exposed on Titan is believed to persist for hundreds to thousands of years — plenty of time for such reactions to take place, similar to processes may have occurred on the early Earth.

Furthermore, six of the other eight planets of our solar system are permanently cold, and hence understanding life's adaptations to cold environments on our planet should be useful in the search for and understanding of life on other planets. Indeed, the current debate about whether life exists on other planets is rooted in understanding what we know about microbial life in cold environments (Rodrigues & Tiedje, 2008).

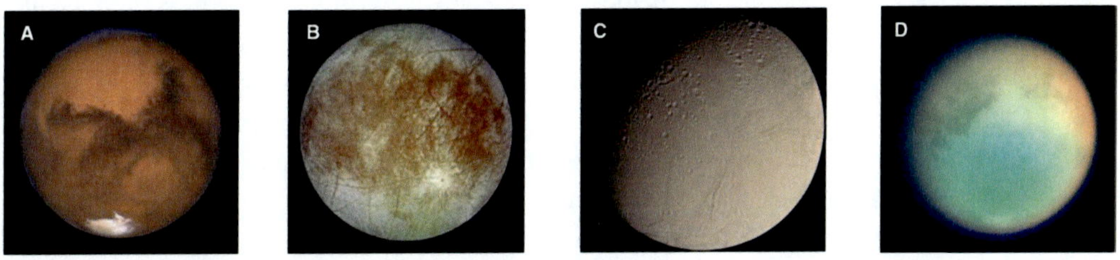

Figure 1. Mars, Europa, Encelado and Titan. There is considerable interest in astrobiology in investigating icy worlds: Mars (A), Europa (B), Enceladus (C) and Titan (D) represent some examples of the main objectives for space research in next missions. (Picture credits: NASA (Mars, Enceladus and Titan) and ESA (Europa)).

The diversity of life on Earth today is a result of the dynamic interplay between genetic opportunities, metabolic capabilities, and environmental changes. For most of their existence, Earth's habitable environments have been dominated by microorganisms and subjected to their metabolism and evolution. As a consequence of geological, climatologic, and microbial processes acting across geological time scales, the physical-chemical environments on Earth have been changing, thereby, determining the path of evolution of subsequent life. Therefore, the co-evolution between organisms and their environment is an intrinsic feature of living systems (Des Marais, 2008).

3. Terrestrial Models of Extraterrestrial Icy Worlds

The origin and evolution of life on Earth and the possibility that life exists outside our planet are very controversial topics. Though global phylogenetic models predict that early organisms were hyperthermophiles, this hypothesis is not universally accepted. It has been pointed out the fact that most molecules with essential biological functions, such as nucleotides, some amino acids, and nucleic acids are thermolabile. Miller's "primitive soup" hypothesis, implying that the first living organisms used RNA as the information molecule, has been challenged by the "metabolic life" theory. This hypothesis suggests that primordial "life" was nothing other than a series of self-catalyzed reactions based in simple compounds such as CO_2 and CO (Rossi *et al.*, 2003). Opinion is divided over two opposing scenarios for the synthesis of prebiotic molecules leading to the origin of microbial life: a hot origin in or near hydrothermal vents in the ocean or a cold origin in veins in ice or on surfaces of mineral grains. The arguments that favor a cold origin of life are attractive. Because of the reduced luminosity of the young Sun, the terrestrial oceans may have been frozen during the early history of Earth (Price, 2009). The ocean temperature on the early Earth depended on the concentration of greenhouse gases such as CO_2 and CH_4 in the atmosphere. If the oceans were initially frozen, CO_2 emitted in volcanic activity would have accumulated in the atmosphere until the greenhouse effect became strong enough to melt the ice. Life could have originated before the frozen oceans melted. Further, several arguments have been given that terrestrial RNA and DNA may have originated in a frozen environment more than 4 billion years ago (Price, 2009).

As a consequence of these controversial theories, terrestrial models of extraterrestrial icy worlds are being intensively studied. On Earth, 90% of oceans have a temperature of 5ºC or less, and when terrestrial habitats are included, over 80% of Earth's biosphere is permanently cold. Among terrestrial environments, 85% of Alaska, 55% of Russia and Canada, 20% of China and the majority of Antarctica are permanently cold (Rodrigues & Tiedje, 2008). Overall, there are three types of environments on Earth where low temperatures are prevalent, namely, ice, cold terrestrial environments and the deep sea. Ice comprises snow, glaciers, frozen lakes, sea-ice and permafrost (Figure 2). Many ecosystems in these regions are enclosed in protected spots such as in aqueous pockets in ice (Thomas & Dieckmann, 2002) and within rocks, where climate is more clement than in exposed areas (Friedmann, 1984).

Examples of cold terrestrial environments include the Dry Valleys of Antarctica and Arctic polar deserts where limit temperature for biological activity is -17ºC. There are a number of records that have demonstrated that some microorganisms can metabolize at temperatures in the range from -17ºC to -20ºC (Siddiqui & Cavicchioli, 2006). Further, temperatures in the oceanic abysses hover around 2ºC at a maximum hydrostatic pressure of 1,100 mars in the Mariana Trench (Marion *et al.*, 2003).

In these environments, life survives and thrives under what seem to be harsh conditions on Earth. While all organisms are composed of nearly identical macromolecules, evolution has enabled such microbes to cope with a broad range of physical and chemical conditions. What are the features that enable some organisms to live under extreme conditions that are lethal to others? The understanding of the versatility of life on Earth, as well as an understanding of the molecular systems that some organisms utilize to survive such extremes will provide a critical foundation for the search for life beyond Earth. These insights will help us understand the molecular adaptations that define the physical and chemical limits of life on Earth and will provide a baseline for developing predictions and hypotheses about life on other worlds (Des Marais, 2008).

Figure 2. Terrestrial models of extraterrestrial icy worlds. Pictures from three icy terrestrial environments: Antarctic permafrost (A), Antarctic glaciers (B), and high mountains in Alps (C).

4. Psychrophiles

As we consider the possibilities of finding life on other planets, it behooves us to evaluate what we know about the limits for life on Earth. In our continued exploration of Earth, we are finding microorganisms in a variety of unexpected habitats. In geothermal hot springs, organisms thriving at temperatures near boiling point of water and at pH values down to 0.5 have been discovered, in the deepest parts of the oceans, those that grow optimally at pressures above 1000 bars and die at pressures below 500 bars, and at the poles those that grow below the freezing point of water and die at temperatures above 10ºC. The discovery of cold-tolerant microorganisms in glaciated and permanently frozen environments has broadened the known range of environmental conditions which support microbial life. These microorganisms, known as psychrophiles are defined as organisms having an optimal temperature for growth at about 15ºC or lower, a maximal temperature for growth at about 20ºC, and a minimal temperature for growth at 0ºC or below. The lowest temperature limit for life seems to be around -20ºC, which is the value reported for bacteria living in permafrost soil and in sea ice. Among these, bacterial communities from Arctic and Antarctic permafrost, subglacial lakes and high mountains are considered representative of these environments in which psychrophilic bacteria are the unique inhabitants.

Figure 3. Psychrophiles. Scanning electron micrographs of ice samples from three Pyrenean glaciers: Monte Perdido (A), Maladeta (B) and Literola (C), illustrating bacterial cells morphology (red arrows) compared with glacial till.

In this chapter, we show the results obtained from some field and laboratory investigations about psychrophiles that live in European glaciers (Figure 3) where they are not only viable but also, in some cases, metabolically active.

5. Biodiversity and Ecology of Psychrophilic Microorganisms

Microbial activity at cold temperatures is restricted to small amounts of unfrozen water inside the permafrost soil or the ice, and to brine channels. Live thrives in these environments with a remarkable microbial biodiversity of mainly bacteria, archaea, fungi (in particular yeasts) and microalgae.

Among the bacteria that have been detected, the most commonly reported microorganisms are the gram-negative α-, β-, and γ-proteobacteria (*Pseudomonas* spp. and *Vibrio* spp.) and the Cytophaga-Flavobacterium-Bacteriodes phylum. Coryneforms, *Arthrobacter* sp. and *Micrococcus* sp. are the most frequently found Gram-positive bacteria. Bacteria generally dominate in number and diversity over Archaea, although in some areas such as deep sea waters, these are found in equivalent numbers. Among identified cyanobacteria, *Oscillatoria*, *Phormidium* and *Nostoc* are dominant in most of the Antarctic habitats (D'Amico *et al.*, 2006). Biodiversity under these extreme conditions has to be considered not only in terms of the number of group subspecies but also in terms of the genes present in the species and the ecosystems they form.

As an example of bacterial biodiversity and ecology in high mountains, in this chapter we show the results obtained from several ice samples from three Pyrenean glaciers which have been studied in order to understand their microbial ecology (Figure 4). We amplified community DNA and constructed clone libraries with Bacteria-specific and Archaea-specific 16s rRNA primers. After removal of potential chimeric sequences and derreplication of identical sequences, phylogenetic analysis demonstrated that around 20 phylotypes could be identified. BLAST analysis revealed high biodiversity in bacterial species and some phylotypes having reduced phylogenetic similarity to currently catalogued species. Some archaeal species were also found (data not shown). Incubation of nutrient-amended melt water at 4°C under several culture conditions resulted in the recovery of viable cells from ice samples. These viable microorganisms, which were also phylogenetically analyzed, raised the question of whether they had been preserved in an inactive state or were able of maintenance their metabolism in ice samples for long periods of time.

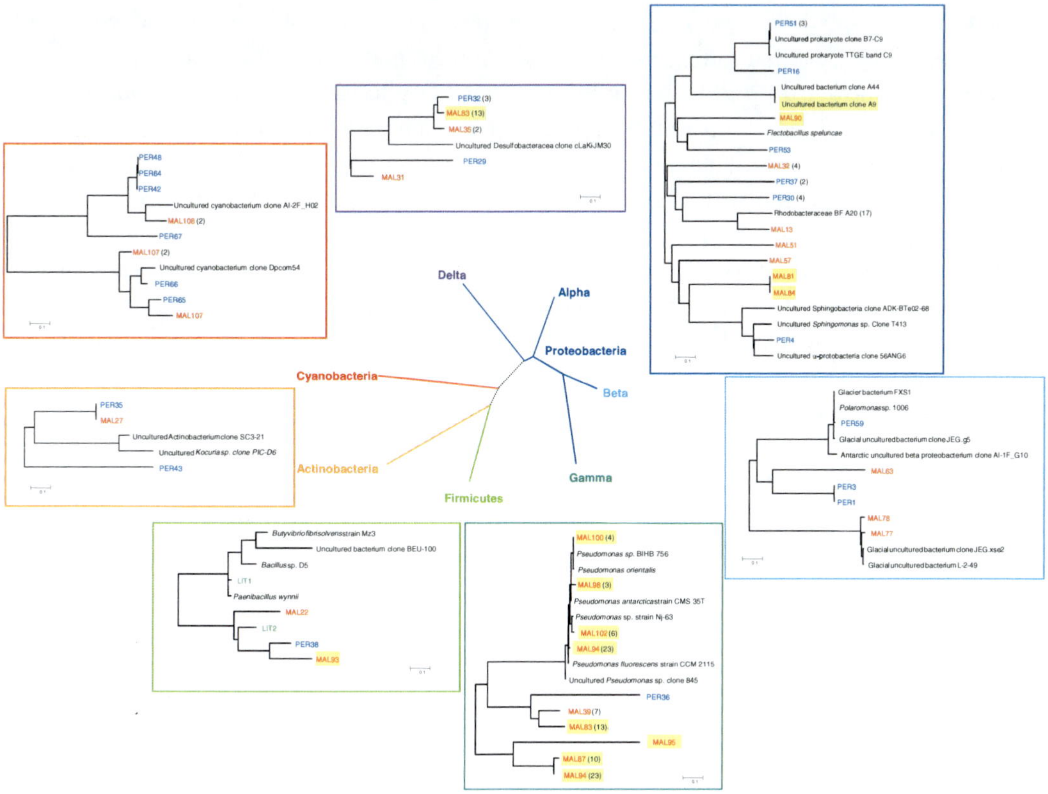

Figure 4. Phylogenetic analysis of encased microbial community DNA in ice samples. A 16S rDNA tree of psychrophiles from three Pyrenean glaciers was constructed in order to understand the microbial composition of these ice samples. The maximum-likelihood phylogenetic tree was performed comparing 16S rDNA phylotypes from ice DNA (blue from Monte Perdido glacier, red from Maladeta glacier or green from Literola glacier) to closely related reference relatives (black). Numbers in parentheses indicate clone frequency. Phylotypes from the meltwater cultures are in yellow boxes. Phylogenetic trees were conducted using the MEGA 4.1 (Beta 3) analysis program (Tamura, et al., 2007).

6. Molecular Mechanisms of Biochemical Adaptation to Cold Environments and Evolution

All of these extremophile organisms are proving that the biochemical machinery of life can be adapted to conditions that, from our anthropocentric perspective appear to be extreme. By studying their molecular adaptations, we could try to identify the critical cellular components that limit life.

As temperature affects essentially every aspect of an organism's physiology, from the basic structure of the macromolecules that are responsible for catalysis and information processing to the rates at which chemical reactions occur, there have been studied different molecular mechanisms of biochemical adaptation to cold environments.

For instance, in their natural habitats, bacteria are frequently exposed to sudden changes in temperature and it has been shown that bacteria use different strategies to cope with temperature changes that are genetically determined and start with registration of the temperature followed by the induction of a subset of genes allowing them to adapt to the stressful situation. These temperature changes are registered by three different thermosensors: DNA, RNA and proteins (Schumann, 2009).

Cold-adapted organisms have successfully evolved features, genotypic and/or phenotypic, to surmount the negative effects of low temperatures and to enable growth in these extreme environments. The ability of psychrophiles to survive and proliferate at low temperatures implies that they have overcome key barriers inherent to permanently cold environments. These challenges include: reduced enzyme activity, decreased membrane fluidity, altered transport of nutrients and waste products, decreased rates of transcription, translation and cell division, protein cold-denaturation, inappropriate protein folding, and intracellular ice formation. Furthermore, genes expressed at low, but not at high temperatures are involved in conjugation and synthesis of flagella among others. In this chapter, we will discuss the current knowledge of these adaptations as gained from extensive biochemical and biophysical studies and also from genomics and proteomics.

6.1. Polyunsaturated Fatty Acids

Low temperatures have an adverse effect on the physical properties of membranes, leading to a reduction in membrane fluidity. The lipid composition governs the physical properties of membranes and it varies with the thermal habitat of microorganisms.

So, psychrophilic microorganisms modify the lipid composition of their membranes to maintain the proper fluidity of the lipid bilayer. This involves the introduction of steric constraints, by reducing the packing order of acyl chains in the membrane. Such steric constraints destabilize the membrane and reduce lipid viscosity. This is achieved for instance by synthesizing lipids with cis-unsaturated double bonds that induce a kink in the acyl chain, or by incorporation of branched lipids or of short fatty-acyl chains that reduce the contacts between adjacent chains (Feller, 2007). This altered composition, which generally do not occur in other prokaryotes, is thought to have a key role in membrane fluidity by introducing steric constraints that change the packing order or reduce the number of interactions in the membrane (D'Amico *et al.*, 2006).

6.2. Cold-Adapted Enzymes

Insight into cold adaptation has occurred through studies performed on analysis of structures, biochemical activity and stability of numerous enzymes from psychrophiles. A range of structural features correlates with enzyme cold adaptation. However, no structural feature is present in all cold-adapted enzymes, and no structural features always correlate with cold adaptation (Siddiqui & Cavicchioli, 2006).

The explosively increasing number of complete genome sequences enables comparative analysis of the general metabolic capacity, and it has been suggested that several novel enzymatic conversions are catalyzed for which no genes are already known, whereas some expected genes for various anticipated metabolic pathways are missing (Rossi *et al.*, 2003). Recently developed proteomic techniques for determining gene function along with differential expression detection and analysis have been applied as high-throughoutput technologies for function discovery and for reconstructing functional networks in psychrophiles. Their ability to proliferate in the cold is predicated on a capacity to synthesize cold-adapted enzymes, which have high specific activities and continue to function at near freezing ambient temperatures, often up to an order of magnitude higher than those observed for their mesophilic counterparts. The commonly accepted hypothesis for this cold adaptation is the activity-stability-flexibility relationship, which suggests that psychrophilic enzymes increase the flexibility of their structure to compensate for the "freezing effect" of cold habitats. This increased flexibility might concern the entire protein or might be restricted to parts of the structure, especially those implicated in catalysis, compared to thermostable homologs (D'Amico *et al.*, 2006).

The intimate linkage between conformational flexibility and protein function carries an important implication for the evolution of protein stability: to function quickly and with accuracy, proteins cannot become too rigid, at least in those regions of the molecule that are involved in recognizing ligands and that undergo changes in conformation during the catalytic cycle. Selection thus prevents proteins from acquiring the highest degree of structural stability that would be possible (Hochachka & Somero, 2002). For instance, some of the main barriers to protein synthesis at low temperatures include reduced activity of transcriptional and translational enzymes, reduced protein folding, and a stabilization of DNA and RNA secondary structures. In psychrophiles, enzymes involved in these processes have adapted to be optimally active at low temperatures.

6.3. Stable Proteins at Cold Temperatures

A cold shock affects multiple levels of cellular physiology such as: (1) decrease in cytoplasmic membrane fluidity affecting membrane-associated functions including active transport and protein secretion, (2) slow or inefficient folding of some proteins, (3) stabilization of secondary structures of RNA and DNA, leading to reduced efficiency of mRNA transcription and translation, and (4) adaptation of ribosomes to the low temperature to function properly (Schumann, 2009).

The exposure of mesophilic organisms to sudden temperature changes, induces the transient over-expression of several proteins -known respectively as heat shock proteins (Hsps) and cold shock proteins (Csps)- that are involved in various cellular processes such as transcription, translation, protein folding and the regulation of membrane fluidity. In particular, increased levels of nucleic-acid-binding proteins (for example, CspA-related proteins) and chaperones, such as GroEL and DnaK, have been frequently reported (Figure 5). However, distinctions do exist between the mesophilic and psychrophilic cold shock response, including the lack of repression of house-keeping protein synthesis and the presence of cold-acclimation proteins (Caps) in psychrophiles. Many of the Csps observed in mesophiles act as Caps in psychrophiles, being constitutively rather than transiently expressed at low temperatures. Further, this differential regulation of expression indicates that a temperature sensory system exists in psychrophiles, and thermosensors at the cell membrane level that sense changes in fluidity have been reported (D'Amico *et al.*, 2006).

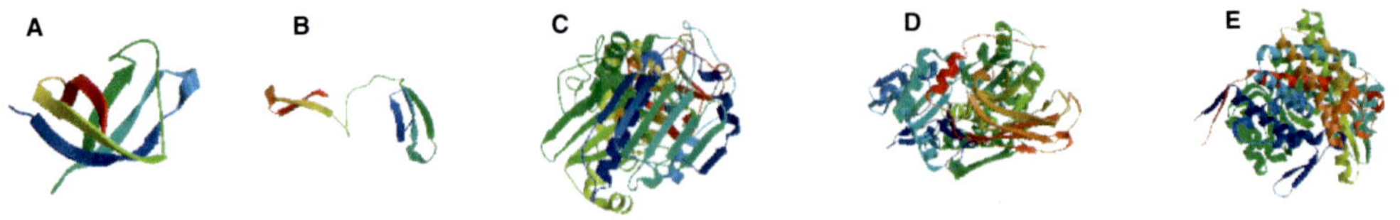

Figure 5. Stable proteins at cold temperatures. Model 3D protein structures of *Shewanella violacea* CspA (accession number Q9S1B7) (A), *S. benthica* KT99 CspD (accession number A9DEF2) (B), *S. oneidensis* HTPG/HSP90 (accession number Q8EFF7) (C), *S. oneidensis* DnaK (accession number Q8EHT7) (D) and *S. oneidensis* GroEL (accession number Q8CX48) (E), obtained with SWISS-MODEL Repository.

In addition, psychrophiles contain antifreeze proteins (AFPs) that have the ability to bind to ice crystals through a large complementary surface and thereby create thermal hysteresis and lower the temperature at which an organism can grow. Other molecules such as trehalose and exopolysaccharides (EPSs) might also have an important role in cryoprotection preventing protein denaturation and aggregation. Furthermore, high concentrations of EPSs modify the physico-chemical environment of bacterial cells, participate en cell adhesion to surfaces and retention of water, favor the sequestration of nutrients, retain and protect extracellular enzymes against cold denaturation and also act as cryopeotectants (Mancuso Nichols *et al.*, 2005).

6.3.1. Heat shock proteins

Molecular chaperones are essential components of the quality control machinery present in the cell (Hartl & Hayer-Hartl, 2002). They optimize bacterial physiological responses, directing client proteins, facilitating developmental signalling and, as Hsps, are indispensable for survival in unpredictable environments. Exposure of cells to cold shock may lead to the induction of molecular chaperones that are also induced by heat shock. There is strong evidence that low temperature, like high temperature can lead to non-native protein structures in vivo and, therefore, to the requirement for enhanced chaperoning activity (Hochachka & Somero, 2002). Hsps have been identified and characterized from various Bacteria and Archaea, and, like most organisms, they generally possess more than one set of Hsps (Laksanalamai & Robb, 2004). Many microorganisms from extreme environments have been well characterized, and increasing access to genomic sequence data has recently allowed the analysis of the protein families related to stress responses. According to these data, Hsps appear to be ubiquitous in extremophiles (Laksanalamai & Robb, 2004). Environmentally sensitive chaperone functions would have different potential consequences for the evolution of populations and lineages under selection in changing environments (Laksanalamai & Robb, 2004; Rutherford, 2003).

6.3.2. Cold shock proteins

When bacteria cultures are shifted from their optimal growth temperature to a lower temperature, cells stop growth immediately and the synthesis of most proteins is repressed. It takes several hours to resume full translation and during this acclimation phase, the synthesis of some proteins increases. These proteins, termed cold shock proteins, are encoded by cold shock genes (Weber & Marahiel, 2003). *E. coli* Csps are essentially constituted by nucleic acid binding proteins involved in different cellular processes like RNA degradation, transcription, DNA replication and supercoiling, translation and ribosome maturation as well as five members of the cold shock protein A (CspA) family which represents the major cold shock response and accumulates up to 10% of the total protein synthesis (Goldstein *et al.*, 1990). CspA and other Csps are among the fastest proteins discovered to fold in response to large temperature changes (Leeson *et al.*, 2000). Besides CspA, *E. coli* codes for two additional proteins CspB and CspG. All three proteins can bind to RNA and destabilize secondary structures thereby acting as RNA chaperones. All these Csps, firstly studied en model organisms as *E. coli* have also been found in many other mesophilic and psychrophilic bacteria (Figures 5A and 5B).

7. Our Latest Results on this Topic in Samples from Icy Environments

The need to predict the functions and structures of new proteins discovered via genomic sequencing has led to the rapidly growing field of extremophile proteomics. A flood of new information has been obtained by the development of appropriate bioinformatic methods for storage and analysis of acquired information in comparative genomics and proteomics studies.

In this chapter, we discuss what we have learned about the role of the Hsps in modifying the lower temperature limit of life and how our studies have provided a new perspective on the functions of these proteins. Among Hsps, we have focused our studies on heat shock protein 90 (HSP90) also known as HTPG (high temperature protein G), which is a conserved molecular chaperone that functions as part of complexes in which different client proteins target it to diverse sets of substrates. In our experiments, HSP90 complexes have been investigated in γ-proteobacteria from mild (*Shewanella oneidensis*) and cold environments (*S. gelidimarina* and *Psychrobacter frigidicola*), to determine changes in HSP90 interactions with client proteins in response to the adaptation to cold environments. Firstly, HSP90 participation in cold adaptation was determined by using the specific inhibitor 17AGG (17-allylamino-geldanamycin) (data not shown). Then, HSP90 was immunoprecipitated from bacterial cultures and the proteins in HSP90 complexes were analyzed by two-dimensional fluorescence Difference Gel Electrophoresis (DIGE), followed by differential in-gel analysis and identified by matrix-assisted laser desorption/ionization-time of flight (MALDI-TOF) mass spectrometry, following a previously described method (Cid *et al.*, 2007). According to HSP90-associated protein analysis (Figure 6), only 15 common proteins were found in both species from the same genus, *S. oneidensis* and *S. gelidimarina*, whereas, a significant higher number of common proteins were found in both psychrophilic species *S. gelidimarina* and *P. frigidicola* 21, ($p<0.001$). Only two HSP90-interacting proteins, the chaperone proteins DnaK and GroEL, were common to the three species. Interestingly, proteins related to energy metabolism (isocitrate lyase, succinyl-CoA synthetase, alcohol dehydrogenase, NAD(+) synthase and malate dehydrogenase) and some translation factors only interacted with HSP90 in psychrophilic bacteria (Figure 6). We can conclude that HSP90 and HSP90-associated proteins might take part in the mechanism of adaptation to cold environments and interestingly, organisms living in similar environments conserve similar potential HSP90 interactors in opposition to phylogenetically related organisms of the same genus, but from different environments.

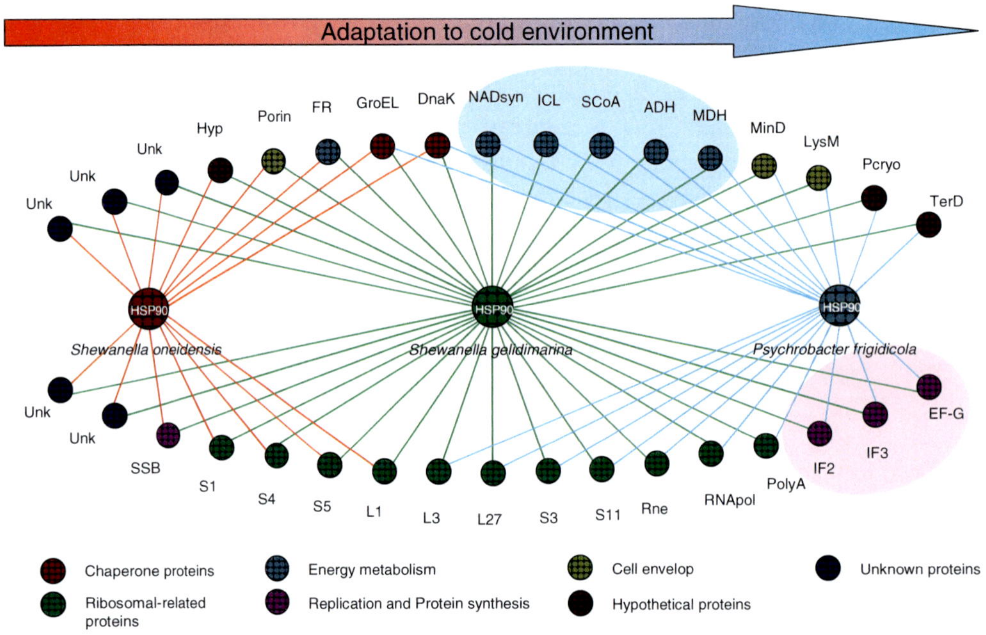

Figure 6. Adaptation to cold environment. Network overview showing the proteins that interact with HSP90 in *S. oneidensis*, *S. gelidimarina* and *P. frigidicola*. HSP90-interacting proteins are colored according to cellular processes. Shaded circles around groups of proteins indicate the main groups of HSP90-interacting proteins found in psycrophilic bacteria: proteins related to energy metabolism (shaded in blue) and protein synthesis (shaded in pink). Abbreviations used: Unk: unknown protein; DnaK: chaperone protein DnaK; FR: chain A, crystal structure of the uncomplexed form of the flavocytochrome reductase; S1: 30 S ribosomal protein S1; GroEL: chaperone protein GroEL; Porin: outer membrane porin; Hyp: hypothetical protein SO4719; SSB: putative single-strand binding protein; L1: 50 S ribosomal protein L1; S4: 30 S ribosomal protein S4; S5: 30 S ribosomal protein S5; Rne: ribonuclease, Rne/Rng family; RNApol: DNA-directed RNA polymerase, subunit beta-prime; IF2: translation initiation factor IF2; polyA: poly (A) polymerase; EF-G: elongation factor G; NADsyn: NAD (+) synthase (glutamine-hydrolyzing); ICL: isocitrate lyase and phosphorylmutase; SCoA: succinyl-CoA synthetase; ADH: alcohol dehydrogenase; MDH: malate dehydrogenase; MinD: septum-site determing MinD; S3: ribosomal protein S3; Pcryo: hypothetical protein Pcryo_1853; TerD: probable tellurium resistance TerD; L3: ribosomal protein L3; IF3: translation initiation factor IF3; LysM: peptidoglycan-binding LysM; S11: 30 S ribosomal protein S11; L27: 50 S ribosomal protein L27.

8. CONCLUSION

- Broadening our knowledge of the range of environments on Earth that are only habitable by some microorganisms and their adaptation to these habitats will be critical for understanding how life might have established itself and survived in habitats beyond Earth. Terrestrial models of extraterrestrial icy worlds must be studied before exploring them in next space research missions.
- The ability of psychrophiles to survive and proliferate at low temperatures implies that they have overcome key barriers and have developed very specific molecular mechanisms of biochemical adaptation which allow them to live in cold environments.
- Because temperature affects essentially every aspect in a microorganism's physiology, cold environments serve as a major driving force in evolution.
- Psychrophiles not only contain special molecules but are also able to modulate their properties such as flexibility and/or their molecular interactions. So, those organisms living in similar environments conserve similar potential protein interactors in opposition to phylogenetically related organisms, that live in different environments.

ACKNOWLEDGMENT

We are indebted to Victor Casado-Lafuente, and Montserrat Martínez-Gomariz for their technical assistance. This research was supported by the Spanish Ministry of Science and Innovation (Grant CTM2008-00304/ANT) and by the Instituto Nacional de Técnica Aeroespacial (INTA).

ABBREVIATIONS

Hsps	heat shock proteins
Csps	cold shock proteins
DnaK	chaperone protein DnaK
GroEL	chaperone protein GroEL
HSP90	heat shock protein 90
HTPG	high temperature protein G.

REFERENCES

Chyba, C. F. (2000). Energy for microbial life on Europa. *Nature, 403,* 381-382.

Chyba, C. F. & Phillips, C. B. (2001). Possible ecosystems and the search for life on Europa. *Proc Natl Acad Sci USA, 98,* 801-804.

Cid, C., Garcia-Bonilla, L., Camafeita, E., Burda, J., Salinas, M., Alcazar, A. (2007). Proteomic characterization of protein phosphatase 1 complexes in ischemia-reperfusion and ischemic tolerance. *Proteomics, 7,* 3207-3218.

D'Amico, S., Collins, T., Marx, J. C., Feller, G., Gerday, C. (2006). Psychrophilic microorganisms: challenges for life. *EMBO reports, 7,* 385-389.

Des Marais, D. J., Allamandola, L. J., Benner, S. A., Boss, A. P., Deamer, D., Falkowski, P. G., Farmer, J. D., Hedges, S. B., Jakosky, B. M., Knoll, A. H., Liskowsky, D. R., Meadows, V. S., Meyer, M. A., Pilcher, C. B., Nealson, K. H., Spormann, A. M., Trent, J. D., Turner, W. W., Woolf, N. J., Yorke, H. W. (2003). The NASA Astrobiology Roadmap. *Astrobiology, 3,* 219-235.

Des Marais, D. J., Nuth, J. A., Allamandola, L. J., Boss, A. P., Hoehler, T. M., Jakosky, B. M., Meadows, V. S., Pohorille, A., Runnegar, B., Spormann, A. M. (2008). The NASA Astrobiology Roadmap. *Astrobiology, 8,* 715-730.

Feller, G. (2007). Life at low temperatures: is disorder the driving force? *Extremophiles, 11,* 211–216.

Friedmann, E. I. & Ocampo-Friedmann, R. (1984). The Antarctic cryoendolithic ecosystem: relevance to exobiology. *Orig. Life, 14,* 771-776.

Goldstein, J., Pollitt, N. S., Inouye, M. (1990). Major cold shock protein of *Escherichia coli. Proc Natl Acad Sci U S A., 87,* 283-287.

Hartl, F. U., & Hayer-Hartl, M. (2002). Molecular chaperones in the cytosol: from nascent chain to folded protein. *Science, 295,* 1852-1858.

Hochachka, P. W. & Somero, G. N. Biochemical adaptation. *Mechanism and process in physical evolution.* New York: Oxford University Press; 2002.

Laksanalamai, P. & Robb, F. T. (2004). Small heat shock proteins from extremophiles: a review. *Extremophiles, 8,* 1-11.

Leeson, D. T., Gai, F., Rodriguez, H. M., Gregoret, L. M., Dyer, R. B. (2000). Protein folding and unfolding on a complex energy landscape. *Proc Natl Acad Sci U S A., 97,* 2527-2532.

Mancuso Nichols, C. A., Guezennec, J., Bowman, J. P. (2005). Bacterial exopolysaccharides from extreme marine environments with special consideration of the southern ocean, sea ice, and deep-sea hydrothermal vents: a review. *Mar Biotechnol, 7,* 253-71.

Marion, G. M., Fritsen, C. H., Eicken, H., Payne, M. C. (2003). The search for life on Europa: limiting environmental factors, potential habitats, and Earth analogues. *Astrobiology, 3,* 785-811.

Price, P. B. (2009). Microbial genesis, life and death in glacial ice. *Can J Microbiol, 55,* 1-11.

Rodrigues, D. F. & Tiedje, J. M. (2008). Coping with our cold planet. *Appl Environ Microbiol, 74,* 1677-1686.

Rossi, M., Ciaramella, M., Cannio, R., Pisani, F. M., Moracci, M., Bartolucci, S. Extremophiles 2002. (2003). *J Bacteriol, 185,* 3683-3689.

Rutherford, S. L. (2003). Between genotype and phenotype: protein chaperones and evolvability. *Nat Rev Genet, 4,* 263-274.

Schumann, W. (2009). Temperature sensors of eubacteria. *Adv Appl Microbiol., 67,* 213-56.

Siddiqui, K. S., Cavicchioli, R. (2006). Cold-adapted enzymes. *Annu Rev Biochem., 75,* 403-433.

Stevenson, D. (2000). Europa's ocean – the case strengthens. *Science, 289,* 1305-1307.

Tamura, K., Dudley, J., Nei, M., Kumar, S. (2007). MEGA4: Molecular Evolutionary Genetics Analysis (MEGA) software version 4.0. *Molecular Biology and Evolution, 24,* 1596-1599.

Thomas, N. D. & Dieckmann, G. S. (2002). Antarctic sea ice –a habitat for extremophiles. *Science, 295,* 641-644.

Weber, M. H. & Marahiel, M. A. (2003). Bacterial cold shock responses. *Sci Prog., 86,* 9-75.

In: Astrobiology: Physical Origin, Biological Evolution... ISBN: 978-1-60741-290-8
Editors: S. Hegedus, J. Csonka, pp. 97-113

Chapter 5

SEEPAGES AND THE ASTROBIOLOGICAL POTENTIAL OF POLAR DUNES ON MARS

A. Kereszturi[1,3*], D. Möhlmann[2], Sz Berczi[1,3], T. Ganti[1], A. Horvath[1,3,5], A. Kuti[4] and A. Sik[1,3]

[1]Collegium Budapest Institute for Advanced Study, H-1041 Budapest, Szentharomsag 2. Hungary
[2]DLR Institute für Planetenforschung, D-12489 Berlin, Rutherfordstr. 2. Germany
[3]Eotvos Lorand University of Sciences, Institute of Geography and Earth Sciences, Planetary Science Research Group, H-1117 Budapest Pázmány 1/c. Hungary
[4]Eotvos Lorand University of Sciences, Department of Astronomy, H-1117 Budapest Pázmány 1/c. Hungary
[5]Konkoly Observatory, H-1525 Budapest Pf. 67. Hungary

ABSTRACT

At the polar dunes of Mars branching dark slope streaks emanating from Dark Dune Spots were observed in local springtime on MOC and HiRISE images. They seep downward with an average speed of 1.4 meter/sol when the temperature of the terrain on 3 km spatial resolution is between 180 and 250 K. Theoretical considerations and calculations suggest interfacial liquid-like water has to be present on Mars, and possibly is causing the observed seepage-like features. The thin water lubricated grains may slide over each other, and/or the stress caused by the refreezing of this interfacial water could also produce such markings, as well as the moving front of solid ice / liquid water phase changes. Such an ephemerally wet environment between the dune grains may have astrobiological significance.

* Corresponding author— E-mail: akos@colbud.hu

INTRODUCTION

In the Collegium Budapest Institute for Advanced Study, the Mars Astrobiology Group works on the possibility of liquid water and life on the red planet. Below, first we summarize the morphological signatures of possible flowing water, and finally astrobiological potential of the dunes, then the theoretical modelling background of the presence of liquid water on Mars today.

There are various seasonal surface markings on the seasonal frost [Christensen et al. 2005; Kieffer, 2003; Malin and, 2000; Piqueux et al. 2003], among them strange spots. The spots have been discovered by [Edget & Malin 1999; Edgett, Supulver, Malin, 2000] and a first proposed explanation for their changes were given by [Bridges et al. 2001]. Dark Dune Spots (DDSs) form a special class of polar seasonal albedo features, whose basic characteristics are the following: they are round shaped; low albedo structures with diameter between 5 and 200 m on the seasonal frost-covered surface of dark dunes; generally show a central darker (umbra-like) core surrounded by a lighter, outer (penumbra-like) ring, although there are many irregular shaped spots also; they appear late winter, grow in size and number, and finally disappear with the disappearance of seasonal frost in early summer. They are present in the southern hemisphere between 65–80 degrees of latitude, and have a yearly reoccurrence to a degree of 50–65%, so about half of the spots appear very close (in meter scale distance) to the location of spots from the previous year. While resembling seasonal dark albedo structures are present in the northern hemisphere, too [Kereszturi et al 2009b], they are excluded in this work because of the differences in their morphology and physical conditions. Their most interesting aspect analyzed in this chapter is the elongated structures on slopes, emanating from them (Figure 1).

METHODS

We analysed HiRISE (Mars Reconnaissance Orbiter, MRO), HRSC (Mars Expess, MEX), MOC (Mars Global Surveyor, MGS) images, topographic data from MGS MOLA PEDR (Precision Experiment Data Record) dataset with processing version L [Smith D. et al. 1999]. Temperature data are from TES (MGS) measurements acquired in nadir-pointing mode [Christensen 1992], using the "vanilla" software, to define annual trend at daytime around 2 pm. These values have spatial resolution of around 3-8 km, so can be taken only as a rough approach of the surface temperature. In this work we surveyed the seasonal albedo changes on dunes inside four craters during southern spring: Russell (54S 12E), two unnamed (68S 2E, 69S 209E), and Richardson crater (72S 180E).

For the analysis to determine the downward direction of slopes, we can use MOLA topography only for large slopes. In most of the cases, the slope direction was identified based on the lighting condition, where the sunward slopes were generally brighter than the opposite ones. Confined slope streaks (see later) were observed only at those locations where the sloping angle was relatively high, and resulted in obvious difference in the brightness relative to the surrounding terrain, because of the illumination.

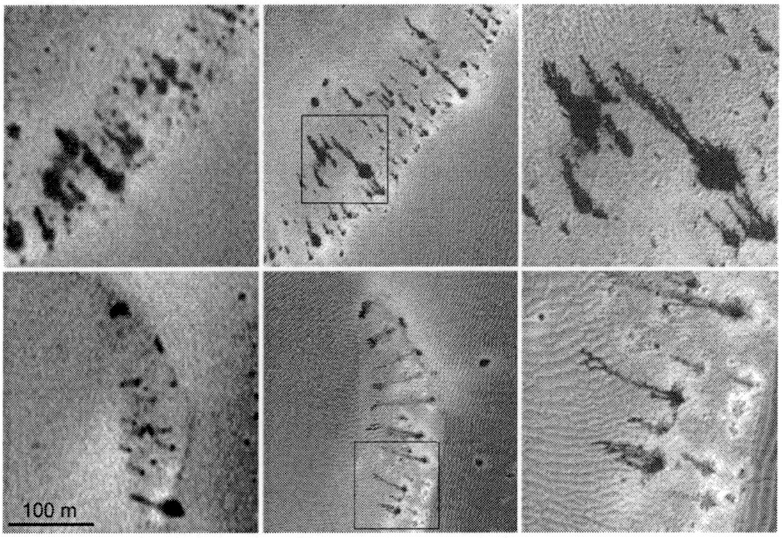

Figure 1. Dark Dune Spots on MOC images (first column R07-00938 acquired at Ls = 220°) and HiRISE (second, third columns, PSP-003609-1110 acquired at Ls = 232°) of an unnamed crater (69S 209E). First and second column images are 300x300 meter wide, while the third column shows the 100x100 meter area marked with the black boxes in the second column. North is up, the Sun illuminates from the left and slopes are tilted toward the left and top left.

We used the following terms to characterize the structures: "slope streak" only for those elongated structures, which stretch in a downward direction on slopes, and emanate from DDSs (in the literature the slope streak term is generally used for low latitude streaks formed possibly by the mass wasting of dust [Aharonson et al. 2003; Beyer et al. 2008; Chuang et al. 2007; Schorghofer et al. 2007]); "movement" was used here for the description of the change in the location of dark-colored surface parts; "flow front" used for the advancing frontal section of these elongated dark slope structures, toward the direction of their movement; "flow speed" means the average speed of the advancement of flow fronts; while "rheologic" for the movement of material on the surface or right below it.

The Possible Seepage Features

Dark slope structures emanating from Dark Dune Spots (DDSs) are visible on Mars [Ganti et al. 2003; Horvath et al. 2001, 2009; Szathmary et al. 2007]. In our previous works [Horvath et al. 2009; Kereszturi A. et al. 2007] we showed that two different phases are present in the DDS development: 1. in early southern spring, fan shaped, diffuse streaks form in downwind direction, possibly by the interaction of wind and geyser-like activity from below the CO_2 ice cover [Kieffer et al. 2006], or solely by the wind blowing away the grains from the surface of spots. In a later phase, with higher temperature and less CO_2 ice cover, water ice is exposed on the surface and confined slope streaks appear.

In this phase carbon dioxide ice does not cover the whole observed area. Unfortunately, the spatial resolution of TES temperature data is not enough to observe the temperature solely at these dark structures. But these dark spots must be warmer than their surroundings. Water ice lies below CO_2 ice in the northern seasonal cap [Bibring et al. 2005; Kieffer and Titus

2001, Schmitt et al. 2005, 2006], and probably a similar situation is present at south, although only smaller water ice patches were observed at the southern polar terrain [Langevin et al. 2006; Titus 2005, 2008], and water ice at the perimeter of the receding annual carbon dioxide cap sometimes tens of kilometers away from the bright permanent cap [Bibring et al. 2004] were identified. Based on this, H_2O ice is probably below the CO_2 ice, and is exposed to the surface at the spots.

Morphology and Appearance

The above-mentioned slope structures analyzed in this chapter were observed on around 10 km diameter dark dune complexes composed of smaller dune units, inside craters at the southern hemisphere. There are small ripples on their top, with width of 5 m–20 m. During the observations, the surface was covered with bright frost (at the beginning CO_2, later possibly somewhere only H_2O). We analyzed the change of around one hundred slope streaks on HiRISE images (Figure 2) and were able to identify the following characteristic sequence of events as the season passes by:

Dark spots appear, somewhere with diffuse, fan-shaped streaks emanating from them, formed probably by sand or dust blown away by wind from the spot and possibly geyser activity.

From these spots confined streaks start to emanate in downward direction on slopes.

The streaks become longer as time passes by, forming a branching pattern. Finally a brighter halo becomes visible around both the spots and the slope streak in some cases, possibly by refreezing phenomenon.

The dark parts become gray and gradually reach the darkening color of the whole dune complex with the advancement of season as ice disappears from the dunes.

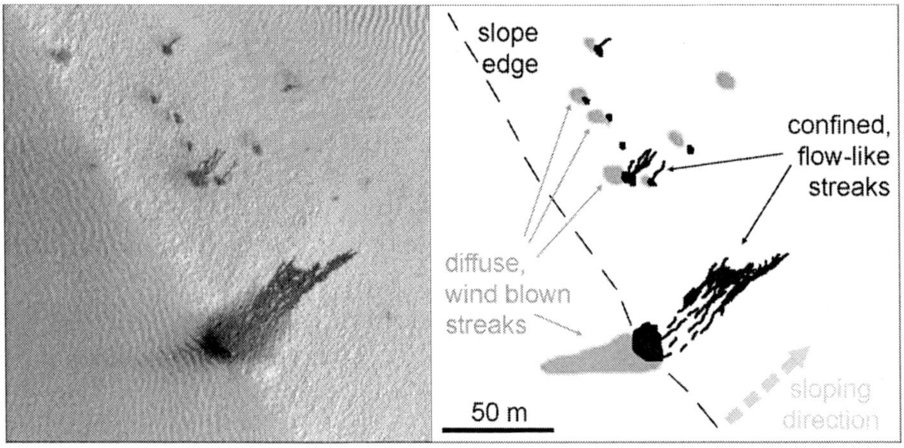

Figure 2. Differences between diffuse wind-blown (gray), and confined flow-like (black) streaks on a 200x200 m section of the HiRISE 3175-1080 image of Richardson crater (left) and its graphical representation (right). The slope's edge runs from the bottom right toward upper left, the slope is tilted toward the upper right and it is brighter because of solar illumination than the darker and nearly horizontal part on the bottom left. North is upward

Slope Streak Movement

The general characteristics of the movement of the confined low albedo slope streaks are:

always start from DDSs downward, no streaks are visible without DDSs. This means that the processes, which trigger the formation of slope streaks are in connection with DDSs,

always follow the sloping direction, regardless to the exposure of the slope faces (e.g. north, south, east, west), suggesting that gravity driven effects dominate the development, opposite to the gas jet process, which forms fan shaped streaks in various directions,

they form a branching network, where the smallest branch has width of 3-6 m, equal to the average distance between the small dune ripples. At many locations individual branches split or coalesce, and during their advancement they follow these small depressions between ripples,

they become longer as time passes by in most (but not all) cases,

at some occasions the source region (the dune spot) becomes lighter during the movement, indicating that the dark material has moved away leaving behind other stuff; or that the source region became covered by whitish frost again; or if only phase changes have caused the dark color, the phase changed back again to the original state,

at the bottom of the slopes, dark pond-like structures form where the streaks arrive at a nearly horizontal surface, stop their movement, and probably accumulate.

There are two subgroups of confined streaks: active and passive streaks. "Active confined streaks" show evident signs of gradual movement of dark flow fronts as the season passes. At the "passive confined streaks", no similar moving flow front was observed, only the small and separated dark parts became visible as the season passed by and finally the separated parts were connected with each other forming a continuous streak – and the passive streak gains similar appearance to an active streak (Figure 3). The elongated morphology suggests the streaks passive today could have been active long ago, possibly during different climate. Hereafter we are dealing with active confined streaks only.

The flow speed values were measured for Martian days (sols) as an average, as if the whole movement had happened during a full Martian sol – though in reality the movement may take place only during a small part of a Martian day. The observed average motion speed was estimated to be up to 1.4 m/sol (Figure 4, 5). At many occasions the dark stuff did not move, or moved slower than 0.1 m/sol. But where the movement was easily observable, it usually happened on meter scale distance on an average Martian day. This suggests the presence of some critical limit above where the movement took place, and than it happens at a reasonable distance.

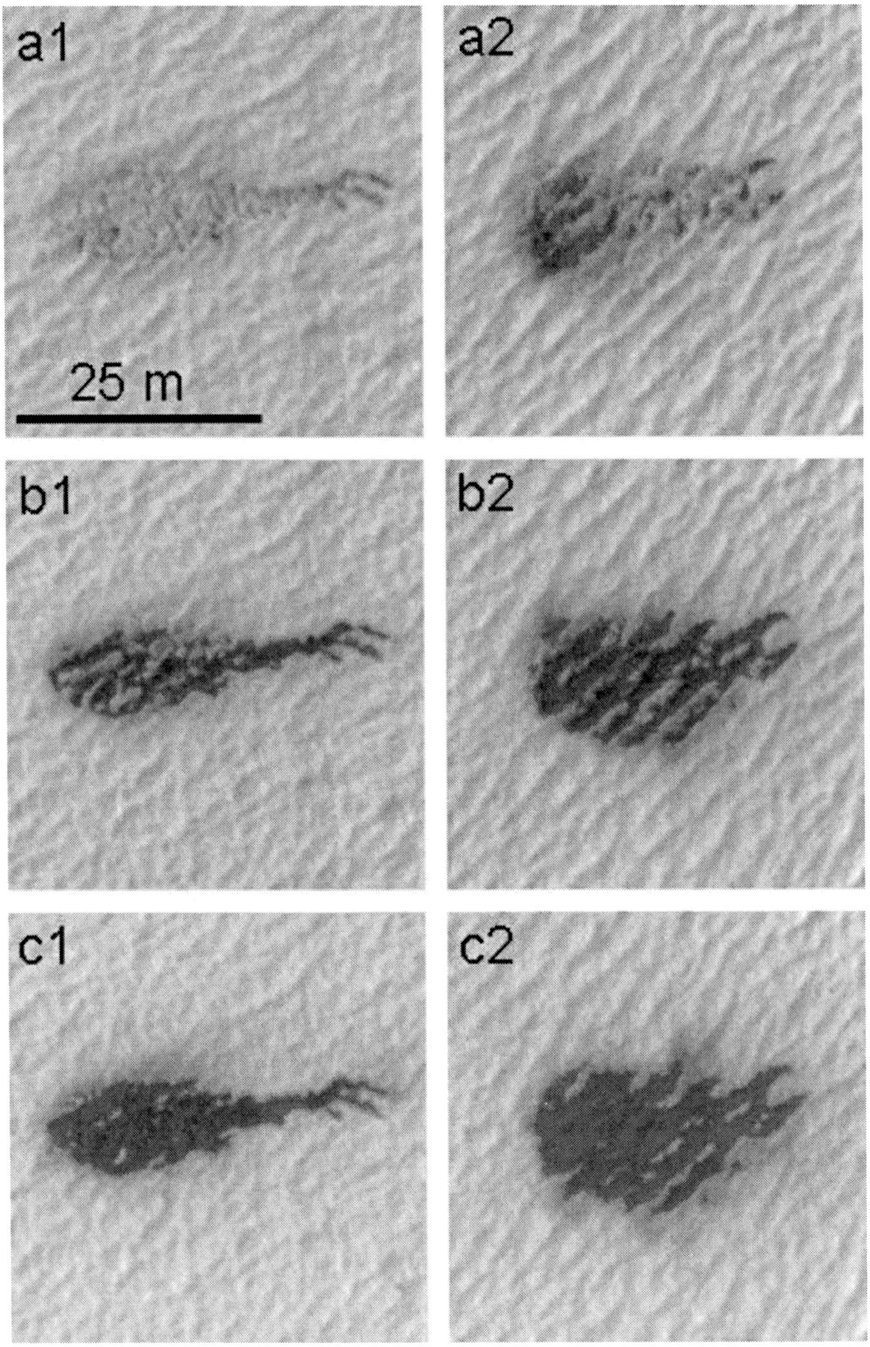

Figure 3. Changes of two passive streaks (left and right column) on 50 × 50 m sections of HiRISE images No. 3432–1115 (a1, a2), 3643–1115 (b1, b2), 3709–1115 (c1, c2), at Ls = 222.9, 233.2, 236.4 respectively. Time passes by downward, and north is upward. The elongated shape toward the right suggests the material in them moved toward the right previously, but currently they show only darkening of small segments, which finally coalesce into a continuous streak (third row).

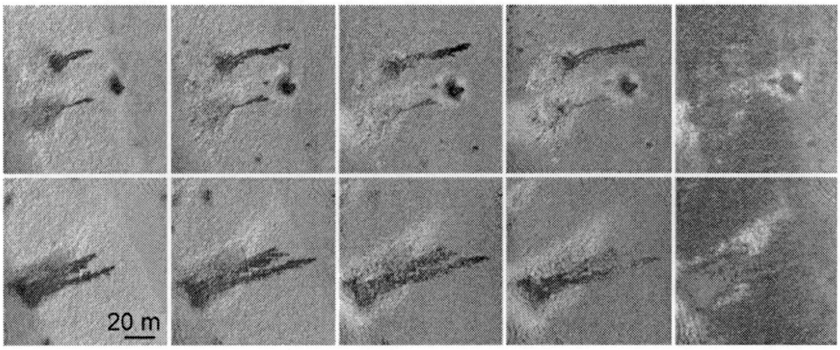

Figure 4. Development of confined slope streaks and Dark Dune Spots in Richardson crater on 100 x 100 meter sized pictures from HiRISE image No. 3175, 3386, 3597, 3742, 3953, which were acquired at Ls = 210.6, 220.7, 230.9, 238.1, 248.5 respectively.

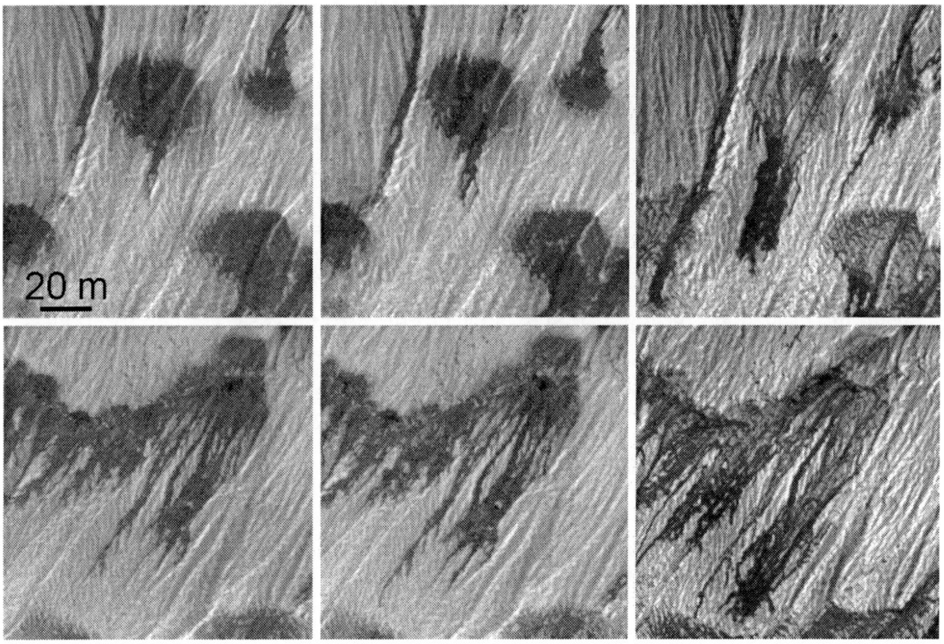

Figure 5. Change of seepage structures emanate from Dark Dune Spots from Russell crater on 100x100 meter images. From left to right HiRISE images No. 2482, 2548, 2904 are visible, which were acquired at Ls = 178.9, 181.9 and 197.9 respectively.

Temperature and Morphology

The four observed craters were situated at 54S, 68S, 69S and 72S latitude. During the springtime warming the average surface temperature reached certain values later in solar longitude for locations farther from the equator. Because of the comparatively low spatial resolution of the detector, it cannot resolve certain parts of the dark features, which may be warmer than the measured "average" value. Despite the low spatial resolution of TES temperature data, the streaks' movement started around the same temperature in the four analysed craters, between 180 K and 200 K.

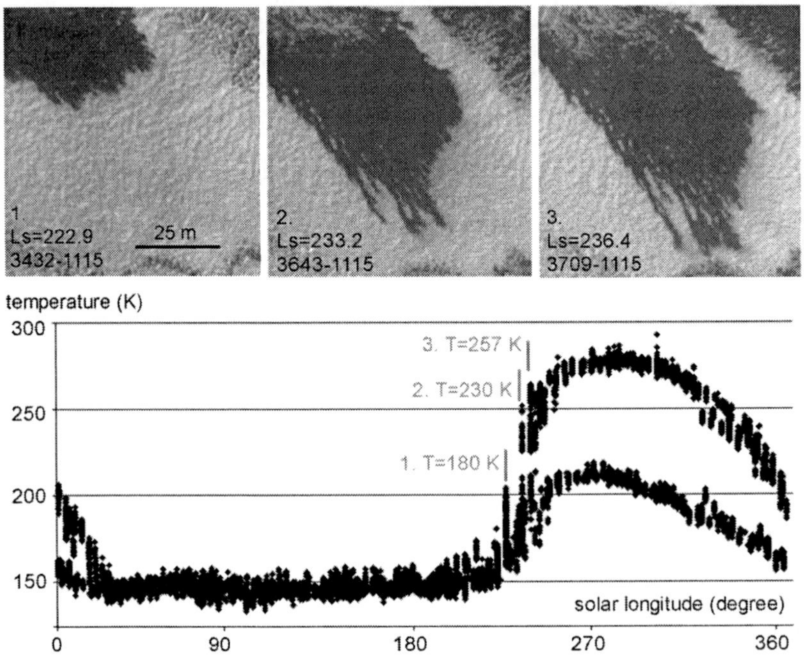

Figure 6. Phases of seepage development (three subsets at the top) and annual temperature curve (bottom) of the unnamed crater at 68S 2E

During the movement of slope streaks, a part of the terrain was free of carbon dioxide ice, as the temperature of the dark slopes is above the CO_2 freezing point (Figure 6.). The flow-like process seems to start when the temperature is above the carbon dioxide frost point, around about 180 - 200 K. The phenomenon called undercooled liquid interfacial water may help in the movements on slopes, as described below.

Interfacial Water on Mars

One of the most important issue in Mars research is the debate on the possibility of liquid water on the surface today. There are models, which suggest that the ephemeral presence of water is limited, but possible on Mars today [Chevrier et al. 2008, 2009; Haberle et al., 2001; Hecht, 2002; Hecht et al. 2009; Malin and Edgett, 2000; Mellon and Phillips, 2001; Motazedian, 2003; Reiss et al. 2004; Renno et al. 2009a; Renno et al. 2009b], while other models contradict this. In the melting sunlight absorbing dark surface may help [Clow 1987], and there are models on the decreasing effect of various salts, which could decrease the melting point of water [Mellon and Phillips, 2001; Knaut and Burt 2002]. Interesting analogs were also discovered from the Antarctica Dry Valleys too for liquid water related structures, resembling to Martian low latitude slope streaks [Head et al. 2007a, 2007b].

In the case of hydrophilic mineral grains embedded in snow/ice, a layer of liquid interfacial water form between the mineral surface and the snow/ice [Möhlmann, 2008], because of the freezing point depression due to the attractive pressure between the surfaces of water-snow/ice and the mineral. This can be described for surfaces of ice (without a mineral counterpart) in terms of "premelting" of ice [Dash et al., 2006] by attractive van der Waals force acting upon the uppermost surface layers of snow/ice (cf. Figure 7.). This attractive van der Waals force F_{vdW}, acting per area F upon the surface of ice of volume V = F d on a mineral surface F is given in case of parallel interfaces by

$$F_{vdW} = \frac{A_{Ilm}}{6\pi d^3} F \tag{1}$$

where A_{Ilm} is the Hamaker constant for the interaction of substances "I" (ice) and "m" (mineral) in presence of interfacial water "l" in between. This force acts also between the molecules of water ice at and very near to the surface of ice, which therefore is prone to "premelting". This configuration is described schematically in Figure 7.

The depression of freezing temperature is related to different acting pressures, as interfacial melting (premelting) via F_{vdW}, via curvature melting in course of surface tension, and pressure melting via $(p_L-p_m)((p_S/p_L)-1)$ by

$$\rho_S q \frac{T_m - T}{T_m} = \frac{A}{6\pi d^3} + \kappa\sigma + (p_L - p_m)\left(\frac{\rho_S}{\rho_L} - 1\right) \tag{2}$$

[cf. Dash et al., 2006, Wettlaufer and Worster, 2006]. where T_m is the standard melting temperature, $\Delta T = T_m - T$ is the freezing point depression, q is for latent heat, q and p for mass density and pressure respectively, the indices L and S indicate the solid ice and the liquid state.

The limiting minimum temperature of the liquid water layer below the bulk melting temperature T_m when van der Waals forces acts only, is given by

$$T = T_m\left(1 - \frac{A}{6\pi q \rho_S d^3}\right). \tag{3}$$

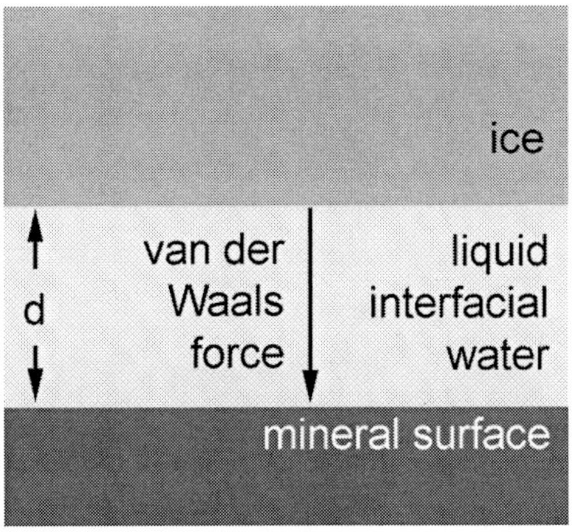

Figure 7. Liquid interfacial water (layer thickness d) between ice and surface of a mineral grain: the "Sandwich model"

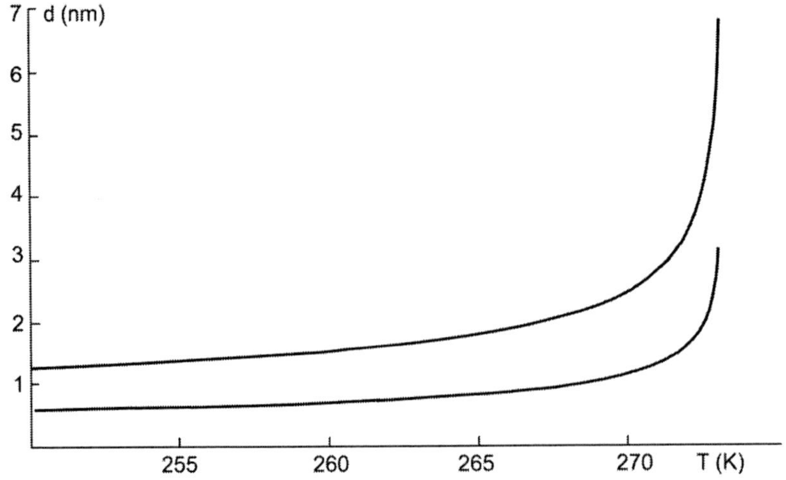

Figure 8. Dependence of the thickness d [m] of the liquid interfacial water layer on the limiting temperature Tl [K] of undercooled water to remain liquid for two Hamaker constant values: A = 10^{-18} J (upper curve) and A = 10^{-19} J (lower curve)

The limiting minimum temperature for interfacial water to remain liquid in dependence on layer thickness and for the range of typical Hamaker constants in nature is shown in Figure 8. It is visible in Figure 9, interfacial water may exist down to about 150 K, as it has experimentally been verified by Person and Derbyshire [1974]. Furthermore, the existence of "unfrozen" water in subzero temperature soils has also been demonstrated and studied by Anderson [1968] and Anderson et al. [1973]. Note that the thickness of the layer of undercooled liquid interfacial water is often measured in film-like "monolayers" (of 0.35 nm), which are to be understood as an orienting average value, not as a stringent geometrical model.

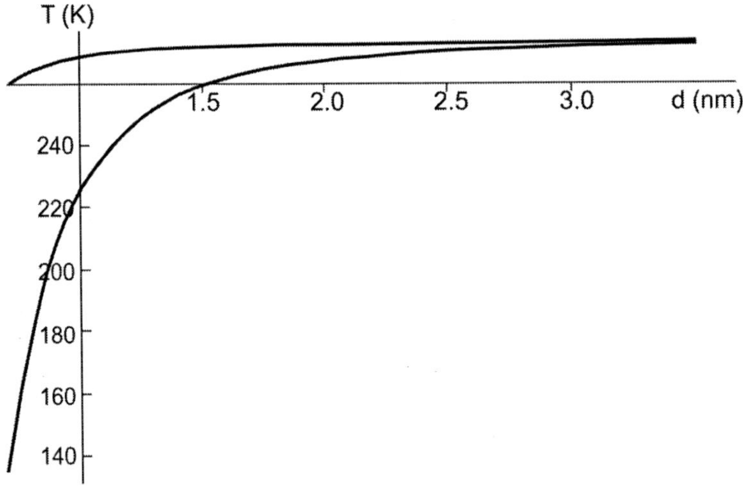

Figure 9. The minimum liquid layer temperature for A=10^{-19} J (upper curve) and A=10^{-18} J (lower curve) in dependence on the layer thickness d. Liquid interfacial water of about two monolayers in thickness may exist down to about 150 K

$$d = \left(\frac{A\,T_m/6\pi}{\rho_S\, q\, \Delta T} \right)^{1/3} \quad (4)$$

Note that the formation of interfacial water from atmospheric humidity will in case of surfaces of salty adsorbents finally lead to salty solutions or brines. This will be followed by an additional freezing point depression, also with the remarkable consequence to grow towards macroscopic amounts of liquid water in the brines (cf. Renno et al., 2009). Insofar, the word "interfacial water" will here to be used for pure as well as for the water in salty aqueous solutions, which are expected to exist on Mars (cf. Chevrier et al., 2008, 2009).

Thus, brines by adding salts, which also tend to increase the thickness d of the layer lower this limiting temperature as shown in Figure 8. and 9. These figures indicate that pure liquid interfacial water must exist on Mars in a temperature band below the bulk melting temperature, and even more also liquid water in brines. The necessary temperatures can be reached at polar latitudes at daytime conditions, in the vicinity of the snow/ice interface above an insolation-heated ground. This temperature can remain "stable" for hours until the ice has left by daytime sublimation or until cooling starts at late afternoon hours (cf. Figure 9.).

Relatively high, near to melting point sub-surface temperatures (with the consequence of possible temporary melting) can be expected in course of effective cooling at the surface of insolated ice/snow (due to sublimation and infrared re-radiation) with heat accumulation in an insolation-absorbing sub-surface, if the thermal conductivity is sufficient small. At least, this will strengthen the possible appearance and amount of interfacial water.

The above-mentioned undercooled interfacial water including brines will be able to cause downhill flow processes on slopes, possibly also lubricating the grains and lowering the force of friction between the grains, and it may seep downward into the deeper subsurface too. This infiltrated water will refreeze in the cooler lower subsurface layers. Since the volume of ice is about 9% larger than the liquid interfacial water, the refreezing must be followed by stresses, that built-up in the upper subsurface, and this may lead to cracks, failures and erosive processes [cf. Fisher 2005] in the upper surface.

LIQUID INTERFACIAL WATER IN SOIL

Liquid interfacial water content of porous soil with specific surface S_M, dry mass m_{dry} and mass M_{H_2O} can simply be estimated via

$$a_m(T,d) = \frac{M_{H_2O}(T,d)}{m_{dry}} = \rho_{H_2O}\, S_M d = S_M\, \rho_{H_2O} \left(\frac{A\,T_m}{6\pi q \rho_S\, \Delta T} \right)^{1/3} . \quad (5)$$

Figure 10 shows measured and modeled (by eq. (5)) soil water contents for four different montmorillonites. Obviously, the content of "nanometric", liquid interfacial water in frozen soil is a macroscopically relevant phenomenon. It may well reach several 10%. This phenomenon is well known as "unfrozen" water in terrestrial permafrost. Analogously to terrestrial permafrost, undercooled liquid interfacial water may appear on Mars preferably in springtime, after the removal of seasonal CO_2-ice cover at sites where water-ice is in contact with mineral grains.

SEEPAGES BY INTERFACIAL WATER

Interfacial water may produce gravitationally driven flows preferably between the translucent covering water ice layer and the warmer ground, which is heated by insolation. The interfacial water will remain near to the melting temperature over longer hours or so, and it must form several monolayers there (cf. Figure 8.). On inclined surfaces as slopes, flows may develop. The thickness of interfacial layer is the greatest at temperatures close to the melting temperature. Depending on grain sizes, for capillary effects in 10 nm range may become sufficient to help wetting and transport process.

On inclined slopes the flow of liquid interfacial water will be counteracted by friction within the liquid layer. The resulted stationary velocity v of a particle with radius R within the liquid layer is given then by

$$v = \frac{2R^2 (\rho_P - \rho_W) g \sin\alpha}{9 \eta} , \qquad (6)$$

where the viscosity η is given by $\eta = 1.38 \times 10^{-3}$ Pa s [Rempel et al., 2001], g is the gravitational acceleration on the surface, α is the inclination of the slope, and ρ_P and ρ_W are the mass densities of the particle and of the water, respectively. Diurnal downslope flow of distances up to about 1 m seem to be possible in the case of the above described slope model (Figure 11.).

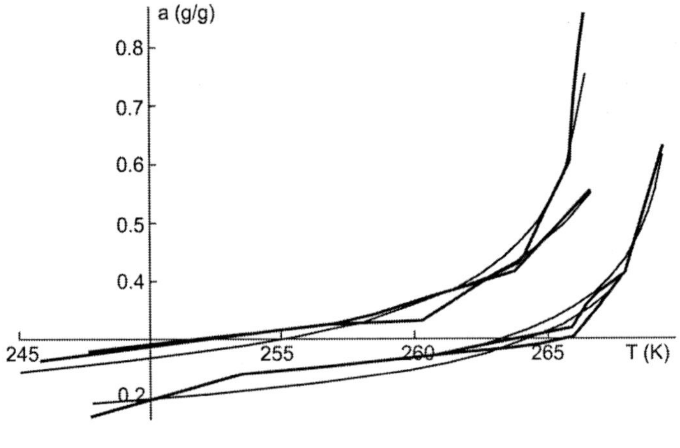

Figure 10. Water contents for four different montmorillonites: Li-, Na-, Ca-, and K-montmorillonites (exp.: thick lines, theor.: thin lines), according to experimental data of Anderson (1968) and by using Eq. (5). (a) Li-montmorillonite reaches up to about 0.8 g/g (uppermost curve), (b) Na-montmorillonite reaches up to about 0.65 g/g (lower right curve), (c) Ca-montmorillonite reaches up to about 0.55 g/g, and (d) K-montmorillonite (very similar to Ca-montmorillonite), b) reaches up to about 0.4 g/g (nearly overlapping curves in the middle)

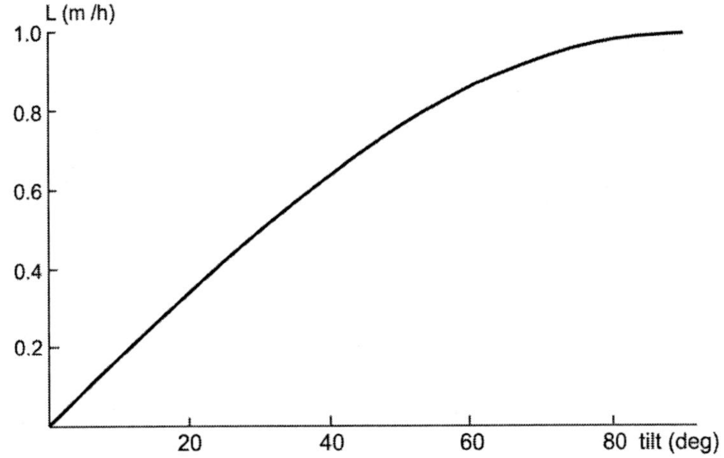

Figure 11. Maximum diurnal downslope flow distance L [m] in dependence on slope inclination ($R = 10^{-5}$ m, $\rho P = 3000$ kgm^{-3}, $\rho W = 1000$ kgm^{-3}).

Possibly related flow-like features on a Martian slope, as observed by HiRISE and studied in the view of the propagation properties at Collegium Budapest [Kereszturi et al. 2008a] are visible in Figure 1., 2., 3., 4. The diurnal progress is about 1 m, but the movements may be active over only a few hours a day. So, the real propagation velocity in the few daily "active hours" is larger. Probably, the presence of liquid interfacial water can also have an influence on its optical properties by solving coloring chemical substances from the solid subsurface materials, what can support the heating process further.

REFREEZING AND EROSION

The refreezing of interfacial water-filled pores will cause erosion. The phase change from liquid to solid results a volume increase (ΔV) of $\Delta V/V = 9\%$ and will cause a pressure or elastic stress in the ice/mineral matrix. This stress can cause failure, and erosion advances in course of refreezing of interfacial water, when the seasonal temperatures approach the local freezing point temperature. With a water ice cover on the surface, seepage and infiltration of interfacial water and its refreezing in the subsurface will happen when the frost point temperature will be reached daytime (in spring) and will decrease below it during nighttime. The erosive stress τ due to refreezing of liquid water can be estimated via

$$\tau = \frac{E}{1-2\nu}\frac{\Delta V}{V} \qquad (8)$$

where E is Young´s modulus, and υ is Poisson´s constant. Typical values for snow and water ice are $E \cong (1\text{-}10)\ 10^9$ Pa, and $E \cong 0.2\text{-}0.5$ given by Mellor (1975). The value $\Delta V/V = 0.09$ indicates that the related stress can cause erosive destruction around 1 GPa. For materials to fail by brittle structure, there is usually a linear correlation between uniaxial compressive strength σ and Young´s modulus with $\sigma /E \cong (2\text{-}3)\ 10^{-3}$ (Mellor, 1975). Compressive stresses for snow are between 1 and 10 MPa (Mellor, 1975). Therefore, refreezing driven stresses can cause erosive destruction in ice/snow mixtures.

CONCLUSION

Narrow, confined and branching dark slope streaks emanating from Dark Dune Spots during local spring were observed on the dunes in four polar craters on Mars. The dark color may be caused by the loss of surface frost cover and/or phase change of the ice at that location. These streaks always start from Dark Dune Spots, stretch and move downward, follow the topographic depressions between the dune ripples. They grow in time, with a speed up to 1.4 meter/sol when the temperature of the terrain on 3-8 km spatial resolution is between 180 and 250 K.

Theoretical considerations and calculations suggest interfacial liquid-like water has to be present on the Martian surface at the solar heated mineral grain – water ice contact, near to the frost point. Such liquid lubricating the grains may help to slide them above each other, and/or water may seep downward on slopes between the ice cover and the grains around 190 K under 10 pr micrometer atmospheric water vapor content. The interfacial water may infiltrate into the subsurface, and freeze to form ice. The stress built up during this process is sufficient to cause erosion.

Such seepage-like process may take place on Mars today and explain the observed streaks. Between the grains, when liquid water is present and the temperature is around 230–250 K, conditions may be favorable for the life process if 1–2 mm thin mineral cover shields again the strong UV radiation but lets enough light in for photosynthesis [Ganti et al. 2009]. The dark dunes based on their color may hold less oxidants than the average reddish surface—so and such microenvironments with interfacial water between the grains have high astrobiological potential [Ganti et al. 2003; Szathmary et al. 2007; Möhlmann 2009; Pocs 2009].

ACKNOWLEDGMENT

This work was supported by the ESA ECS-project No. 98076 and the Hungarian Space Office.

REFERENCES

Aharonson, O., Schorghofer N. & Gerstell F. M. (2003). Slope streak formation and dust deposition rates on Mars. *J. of Geophys. Res., 108*, E12, 5138, doi:10.1029/2003JE002123.

Bridges, N. T., Herkenhoff, K. E., Titus, N. T. & Kieffer, H. H. (2001). Ephemeral dark spots associated with martian gullies. *Lunar and Planet. Sci. Conf., 32*. abstract 2126.

Chevrier, V. F. & Altheide, T. S. (2008). Low temperature aqueous ferric sulfate solutions on the surface of Mars. *Geophys. Res.* Lett., *35*, L22101, doi:10.1029/2008GL035489.

Chevrier, V. F., Hanley, J. & Altheide, T. S. (2009). Stability of perchlorate hydrates and their liquid solutions at the Phoenix landing site. Mars, *Geophysical Research Letters, 36*, LXXXXX, doi:10.1029/2009GL037497.

Christensen, P. &erson, D. L., Chase, S. C., Clark, R. N., Kieffer, H. H., Malin, M. C., Pearl, J. C., Carpenter, J. B. & Nuno, B. F. (1992). Thermal Emission Spectrometer Experiment: Mars Observer Mission. *J. of Goephys. Res.*, *97*, 7719-7734.

Christensen, P. R., Kieffer, H. H., Titus, T. N. (2005). Infrared and Visible Observations of South Polar Spots and Fans. *American Geophysical Union, Fall Meeting* abstract P23C-04.

Clow, G. D. (1987). Generation of liquid water on Mars through the melting of a dusty snowpack. *Icarus, 72*, 95-127.

Edgett, K. S. & Malin, M. C. (1999). MGS MOC the first year: sedimentary materials and relationships, *Lunar and Planet. Sci. Conf., 30,* abstract 1029.

Edgett, K. S., Supulver, K. D. & Malin, M. C. (2000). Spring defrosting of martian polar regions: Mars Global Surveyor MOC and TES monitoring of the Richardson crater dune field, 1999–2000. *Mars Polar Science Exploration, 2,* abstract 4041.

Fisher D. A. (2005). A process to make massive ice in the martian regolith using long-term diffusion and thermal cracking, *Icarus, 179*, 387-397.

Ganti, T., Berczi, Sz., Horvath, A., Kereszturi, A., Pócs, T., Sik, A. & Szathmary, E. (2006). Hypothetical time sequence of the morphological changes in global and local levels of the dark dune spots in polar regions of Mars, *Lunar and Planet. Sci. Conf.,* 37, st. abstract 1918.

Ganti, T., Horvath, A., Berczi, Sz., Gesztesi, A. & Szathmary, E. (2003). Dark Dune Spots: Possible Biomarkers on Mars? *Origins of Life and Evolution of the Biosphere, 33*, 515-557.

Ganti T., Pócs T., Berczi Sz., Horvath A., Kereszturi A., Sik A. & Szathmary E. (2009). Ideal microhabitats on Mars: the astrobiological potential of polar dunes. *Lunar and Planet. Sci. Conf., 40,* abstract 1618.

Haberle, R. M., McKay, C. P., Schaeffer, J., Cabro, N. A., Grin, E. A., Zent A. P., Quinn, R. (2001). On the possibility of liquid water on present-day Mars. *J. Gephys. Res., 106*, E10, 23317-23326.

Hamaker, H. C. (1937). The London-van der Waals Attraction between Spherical Particles, *Physica* IV, No., *10*, 1058.

Head, J. W., Marchant, D. R., Dickson, J. L., Levy, J. S. & Morgan, G. A. (2007a). Slope streaks in the Antarctic Dry Valleys: characteristics, candidate formation mechanisms, and implications for slope streak formation in the Martian environment. *International Conf. on Mars, 7,* abstract 3114.

Head, J. W., Marchant, D. R., Dickson, J. L., Levy, J. S., Morgan, G. A., Kreslavsky, M. (2007b). Mars gully analogs in the Antarctic Dry Valleys: geological settings and processes. *International Conf. on Mars, 7,* abstract 3118.

Hecht, M. H. (2002). Metastability of Liquid Water on Mars. *Icarus, 156*, 373-386.

Hecht, M. H., Catling, D. C., Clark, B. C., Deflores, L., Gospodinova, K., Kapit, J., Kounaves, S. P., Ming, D. W., Quinn, R. C., West, S. J. & Young, S. M. M. (2009).

Perchlorate in Martian Soil: Evidence and Implications. *Lunar and Planet. Sci. Conf, 40,* abstract 2420

Horvath, A., Ganti, T., Gesztesi, A., Berczi, S. z. & Szathmary, E. (2001). Probable evidences of recent biological activity on Mars: appearance and growing of dark dune spots in the south polar region. *Lunar and Planet. Sci. Conf, 32,* abstract 1543.

Horvath, A., Kereszturi, Á., Berczi, S. z., Sik, A., Pócs, T., Gesztesi, A., Ganti, T. & Szathmary, E. (2005). Annual change of Martian DDS-seepages. *Lunar and Planet. Sci. Conf, 35,* abstract 1128.

Horvath, A., Kereszturi, Á., Berczi, Sz., Sik, A., Pocs, T., Ganti, T., Szathmary, E. (2009). Analysis of dark albedo features on a Southern Polar dune field of Mars. *Astrobiology, 1.*, 90-103.

Kereszturi, A., Möhlmann, D., Berczi, Sz., Horvath, A., Ganti, T., Kuti, A., Pocs, T., Sik, A. & Szathmary, E. (2008a). Analysis of possible interfacial water driven seepages on Mars. *Lunar and Planet. Sci, Conf., 39th* abstract 1555.

Kereszturi, A., Sik, A., Horvath, D., Reiss, R. & Jaumann, Neukum G. (2007). Season-dependent behavior of Dark Dune Spots on Mars. *Lunar and Planet. Sci. Conf, 38,* abstract 1864.

Kereszturi, A., Berczi, S. z., Horvath, A., Ganti, T., Kuti, A., Pocs, T., Sik, A. & Szathmary, E. (2008b). North-south comparison of springtime dark slope structures on Mars, and the possibility of liquid water. *European Planetary Sci. Conf, 3,* EPSC2008-A-00386.

Kieffer, H. H. (2003). Behavior of Solid CO_2 on Mars: A Real Zoo. *International Conf. on Mars, 6,* abstract 3158.

Kieffer, H. H., Christensen, P. R. & Titus, T. N. (2006). CO_2 jets formed by sublimation beneath translucent slab ice in Mars' seasonal south polar ice cap. *Nature, 442*, 793-796.

Knauth, L. P., Burt, D. M. (2002). Eutectic Brines on Mars: Origin and Possible Relation to Young Seepage Features. *Icarus, 158*, 267-271.

Malin, M. C. & Edgett, K. S. (2000). Frosting and defrosting of martian polar dunes, *Lunar and Planet. Sci. Conf, 31,* abstract 1056.

Mellon, M. (1975). A review of basic snow mechanics. *Snow Mechanics symposium* 251, IAHS Publ. *114*.

Mellon, M. T., Phillips, R. J. (2001). Recent gullies on Mars and the source of liquid water. *J. Geophys. Res., 106*, E10, 23165-23179.

Motazedian, T. (2003). Currently Flowing Water on Mars. *Lunar and Planet. Sci. Conf, 34,* abstract 1840.

Möhlmann, D. (2008). The influence of van der Waals forces on the state of water in the shallow subsurface of Mars. *Icarus, 195,* 131-139.

Möhlmann D. (2009). Are nanometric films of liquid undercooled interfacial water bio-relevant? *Cryobiology*, doi:10.1016/j.cryobiol.2009.01.004, in print.

Piqueux, S., Byrbe, S. & Richardson, M. I. (2003). Sublimation of Mars's Southern seasonal CO_2 ice cap and the formation of spiders. *J. Geophys. Res., 108*, 5084.

Pocs, T. (2009). Cyanobacterial crust types, as strategies for survival in extreme habitats. *Acta Botanica Hungarica, 51(1-2),* 147-178.

Reiss, D., van Gasselt, S., Neukum, G. & Jaumann, R. (2004). Absolute dune ages and implications for the time of formation of gullies in Nirgal Vallis, Mars. *Lunar and Planet. Sci., Conf, 35,* abstract 1639.

Renno, N. O., Bos, B. J., Catling, D., Clark, B. C., Drube, L., Fisher, D., Goetz, W., Hviid, S. F., Keller, H. U., Kok, J. F., Kounaves, S. P., Leer, K., Lemmon, M., Madsen, M. B., Markiewicz, W. J., Marshall, J., McKay, C., Mehta, M., Smith, M., Smith, P. H., Stoker, C., Young, S. M. M. & Zent, A. (2009a). Physical and Thermodynamical Evidence for Liquid Water on Mars. *Lunar and Planet. Sci. Conf, 40,* abstract 1440.

Renno, N. O., Bos, B. J., Catling, D., Clark, B. C., Drube, L., Fisher, D., Goetz, W., Hviid, S. F., Keller, H., Kok, J. F., Kounaves, S. P., Leer, K., Lemmon, M., Madsen, M., Bo, Markiewicz, W., Marshall, J., McKay, C., Mehta, M., Smith, M., Zorzano, M. P., Smith, P. H., Stoker, C. & Young, S. M. M. (2009b). Physical and Thermodynamical Evidence for Liquid Water on Mars. *Lunar and Planet. Sci. Conf, 40,* abstract 1440.

Ryan, J. A. & Sherman, R. D. (1981). H_2O Frost Point Detection on Mars? *J. Geophys. Res., 86,* No. C1, 503-511.

Smith, D. et al. (1999). NASA PDS, Washington Univ., St. Louis, Mo.

Szathmáry, E. (2005). Signs of water runoff and its relation to possible living organisms on Mars, *Astrobiology: Life in extreme conditions conference*, Blindern, Oslo, Norway.

Szathmary, E., Ganti, T., Pocs, T., Horvath A., Kereszturi, A., Berczi, S. z. & Sik, A. (2007). Life in the dark dune spots of Mars: a testable hypothesis - in *Planetary Systems and the Origin of Life*, ed. Ralph Pudritz, Paul Higgs, Jonathan Stone, Cambridge University Press, 241-258.

Wettlaufer, J. D. & Worster, M. G. (2006). Premelting Dynamics, *Annual Reviews, 38*, 427-452.

In: Astrobiology: Physical Origin, Biological Evolution... ISBN: 978-1-60741-290-8
Editors: S. Hegedus, J. Csonka, pp. 115-130

Chapter 6

PREHISTORY OF THE INITIAL MICROBIAL COMMUNITIES: THE WAY FROM ABIOTIC ORGANIC MICROSYSTEMS TO PROKARYOTIC CELLS

Vladimir Kompanichenko
Institute for Complex Analysis, 4, Sholom-Aleyhem Street,
Birobidzhan 679016, Russia.

ABSTRACT

This chapter is devoted to consideration of the method of the transition from prebiotic organic microsystems to the primordial communities of prokaryotes. During the initial transformations prebiotic organic microsystems acquired the unstable bifurcate state, with its following relative stabilization by means of balanced oscillations around the bifurcation point. Then microsystems of the type converted into the primary living units – probionts, which were characterized with the ability to concentrate free energy and information already. Through the synergy effect probionts were organized into the communities that represented themselves minimum self-sufficient unit of life. The appeared communities of probionts evolved to more complex communities of progenotes (characterized by C.Woese) and then prokaryotes. This earliest biological evolution proceeded through development of the initial nucleoprotein interaction, formation of RNA and then DNA, etc. Emergence of species had led to conversion of the initial communities of probionts and progenotes ("embryos" of ecosystems) into the real ecological systems, which were hierarchically organized of different microbial species.

INTRODUCTION

Investigation of the prehistory of initial microbial population on our planet is important to understand further evolution of the biosphere. The basic effort in this way was done by C. Woese (1987). He theoretically reconstructed *the progenote* – a type of the extinct organism that was considered as the universal ancestor for Archaea and Bacteria. However, the earlier prehistory of the microbial world goes into uncertainty because a method of the transformation of abiotic organic microsystems into simplest living units is unclear. The attempt to clarify this method is made in the framework of the systemic conception of the origin of life (Kompanichenko, 2004, 2008, 2009). The systemic difference between an assembly of organic matter (abiotic system) and a living unit (biotic system) is expressed in terms of the four key biological properties: a) the ability to concentrate free energy and information (by means of their extraction from the environment); b) the ability to exhibit an intensified counteraction to external influences; c) expedient behavior (or expedient character of interaction with the environment); d) regular self-renovation at different hierarchical levels (from molecular to ecosystemic), including self-reproduction. In accordance with the conception, these properties first appeared in the simplest living units – probionts. Further evolution of probionts up to prokaryotes was not reconstructed in the mentioned articles. The aim of this work is to consider briefly the entire way from organic microsystems to the present prokaryotes being at the root of the Phylogenetic Tree. The proposed theoretical reconstruction includes four intermediate units between them: (a) a prebiotic microsystem at the unstable (bifurcate) state; (b) pulsing (bistate) prebiotic microsystem; (c) probiont; (d) progenote.

Prebiotic Organic Microsystems

Several prebiotic organic microsystems were proposed by various researchers: proteinoide microspheres (Fox, Dose, 1977; Fox et al, 1996), RNA-World macromolecules (Gilbert, 1986; Joice & Orgel, 1993), liposomes (Deamer, 1986, 2004; Luisi, 2000), aromatic hydrocarbons (Ehrenfreund et al, 2006). The prebiotic models demonstrate some signs of activity relevant to the vital processes. For instance, a coacervate is able to grow and selectively extract substance from the environment, a catalytic activity peculiar to proteinoide microspheres, some RNA macromolecules are able to self-replicate. However, although these models displayed the ability for chemical complication during the conducted experiments, they have not been converted into simplest living unit by now. It seems that the principal problem consists in their inability to force the initial signs of activity up to the selfmaintaining regime, giving rise to metabolism, regular self-replication and other vital dynamic processes. The all-round consideration of this problem in the framework of the systemic conception allowed formulate the following thesis: transformation of a prebiotic organic microsystem into a living unit is possible only at the state of nonequilibrium bifurcate transition and under oscillating conditions in the outside world (Kompanichenko, 2003, 2004, 2008, 2009).

Behaviour of the biologically important organic molecules and the suggested prebiotic models is still not experimentally explored under nonequilibrium oscillating conditions: almost all experiments on prebiotic chemistry have been carried out in stable conditions. So, it is unclear which of the models, or their combination, was the most suitable to be transformed into initial forms of life on the early Earth. Due to this reason, it is accepted in this chapter the conditional composition of a prebiotic microsystem: mixture of the biologically important organic molecules (aromatic hydrocarbons, lipids, amino acids, nucleic acids) and/or their precursors. The optimum ratio between them can be defined further in the course of the experiments on prebiotic chemistry under oscillating far-from-equilibrium conditions.

The minimum temperature in the geological cradle of life can be evaluated as 70ºC at the time of the origin of life on Earth (3.8-4.4 Ga). This magnitude characterizes seawater temperature of the Hadean ocean. The oxygen isotopic record in marine cherts shows the temperature of the Hadean ocean +80ºC at age 3.8 billion years ago (Ga) and +70ºC (±15ºC) at age 3.5 Ga (Knauth, 1992; Knauth & Lowe, 2003; Schwartzman, 1999). Similar result was obtained on the basis of the exploration of the silicon isotopic compositions (Chaussidon & Franquis, 2006). The oldest records of life are about 3.5 Ga, in accordance with the exploration of microfossils, and 3.85 Ga on the basis of biogeochemical data (Schidlowski, 1997; Schopf et al, 2002; Westall et al, 2006). The maximum temperature limit in the geological cradle of life is uncertain because a temperature of water in hydrothermal systems and their submarine discharges can be much higher than water boiling point, up to 300-400ºC. The upper temperature limit for some present Archaea is 121ºC. These data allow us to agree with the microbiologists, who place thermophilic or hyperthermophiic forms Archaea and Bacteria at the root of the Phylogenetic Tree (Stetter, 1995; Woese, 1987; Xue et al, 2003). Of course, emergence of some thermophilic species from mesophilic forms during the following biological evolution seems to be possible.

High-temperature and wet conditions are appropriate for the listed organic compounds or their assemblies to exist. In particular, we found some aromatic hydrocarbons (naphthalene, biphenyl, fluorene, phenathrene, etc.) in the sterile water-steam mixture with the temperature 175ºC from the borehole in Kamchatka geothermal region. In this region Mukhin et al (1979) discovered glycine in the sterile water-steam mixture, and several other amino acids in the inhabited hot spring. Fatty acids, their precursors and liposomes are usual components of high temperature fluid. Upper temperature limit for the synthesis of nucleotides is 50-60ºC, as the exception up to 90 ºC. So, in principle, prebiotic organic microsystems could be composed of all biologically important organic compounds in the high temperature medium (70 ºC and higher) (Holm & Andersson, 2005; Russell, 2003; Schwartzman, 1999).

The requirement of nonequilibrium conditions connects the origin of life process with the wide range of the specific events related with nonequilibrium bifurcate transitions and formation of dissipative structures. The broad class of dissipative structures embraces as some specific non-living systems (oscillating chemical reactions, Benard cells, etc.) as all living organizations (Nicolis & Prigogine, 1977; Prigogine & Stengers, 1984).

Prebiotic Microsystems at the Unstable Bifurcate State

There exist two principal mechanisms of evolution of natural systems. The first one consists in gradual accumulation of small changes in a system. Importance of such mechanism for the biological evolution was emphasized C. Darwin. The second mechanism is revolutionary and leads to the radical transformation of a system. This mechanism is a bifurcation. The mechanism acts through the bifurcate transition of a system under nonequilibrium conditions. A bifurcation occurs when a system cannot exist further in the current conditions due to changes that have occurred in the surroundings or inner mutations. The universal scheme of bifurcate transition of a system is the following: stable existence of a system → rise of instability through powerful fluctuations → the highest point of instability (point of bifurcation, or critical point), radical change of the system's structure → choice of new path of development → the next period of its stable existence (Figure 1 (a)). At the bifurcation point a system undergoes many accidental changes that may influence the choice of its further development path. This is the reason why potential paths of a system's development bifurcate at the moment of its highest instability. Finally, the system chooses one of many permissible ways, which can be united into two principal trends: (1) complication through self-organization (Trend A), (2) simplification and degradation (Trend B), up to full destruction (Trend B'). A reverse change of the external conditions initiates the back transition of a system into the initial state in accordance with the Trend C (Figure 1 (b)).

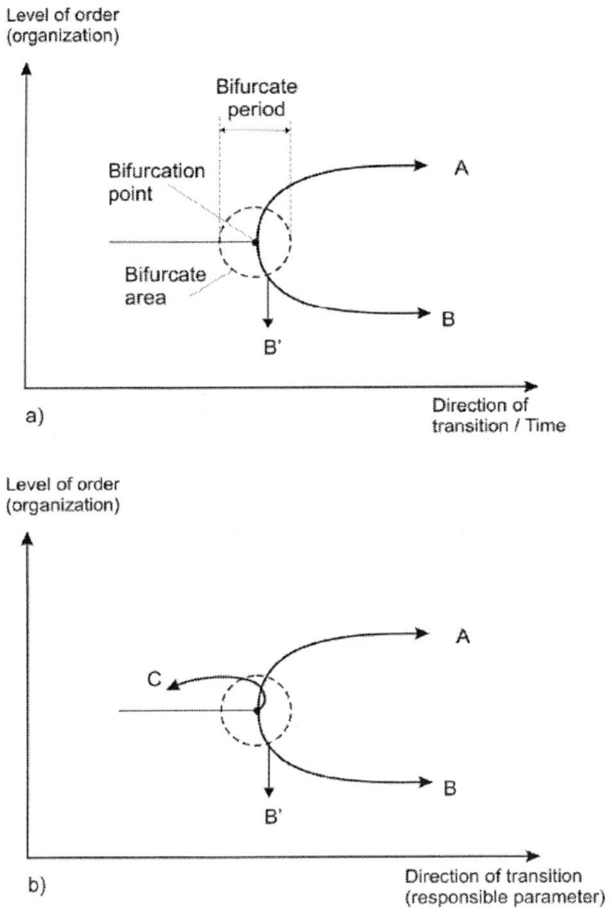

Figure 1. Principal scheme of bifurcate transition of a natural system under nonequilibrium conditions: a – direct transition from the initial stable state into one of permissible advanced stable states due to changed conditions in the outside world;. b – direct and reverse transitions (in case oscillating conditions in the outside world). A –trend to advanced higher-organized state; B – trend to advanced lower-organized state; B' – trend to complete destruction; C - reverse trend to the approximately initial state.

The mechanism of radical transformations of natural systems under far-from-equilibrium conditions is well-explored in the framework of the theory of dissipative structures and synergetics (Haken, 1978, 2003; Nicolis & Prigogine, 1977; Prigigine, 2003; Prigogine &

Stengers, 1984). Behaviour of a chemical system under nonequilibrium conditions radically differs from its behaviour under conditions near equilibrium. Around the bifurcation point a chemical system acquires some specific properties, which are not peculiar to it during a period of the stable existence. The most essential of these bifurcate (or critical) properties are the following: (1) sharp heterogeneity with the intensive counter processes along and against the gradients; (2) continuous fluctuations (waves) and re-arrangement of molecules; (3) integrity through cooperative processes; (4) incessant exchange by matter and energy with the surroundings (open system). Let us consider the obvious analogies between these properties and the corresponding characteristics of living organisms. A comparison is drawn in the Table 1. Each of four main critical properties is distinctly reflected in the vital processes. Although the corresponding characteristics of living cells do not express the unique essence of life, they are in its background. Loss of any of these characteristics would make life impossible. Meanwhile, the critical properties are temporal characteristics of a chemical system: they appear at the beginning of the bifurcate transformation and disappear with the completed transition to the advanced stable state. On the basis of these facts the following thesis is formulated: *Life on Earth arose from the bifurcate state* that was inherent to still indeterminate kind of organic microsystems.

Table 1. Correlation between the principal properties of a chemical system at the bifurcate state and the corresponding characteristics of a living organism

Main properties of a chemical system at the bifurcate state	The corresponding characteristics of a living cell
1. Sharp heterogeneity with the intensive counter processes along and against the gradients. Multitude high gradients (concentration, pressure, temperature, electric field, and other kinds) appear throughout in a chemical system that approaches the critical point of transition. Its structure becomes highly heterogeneous. Values of the gradients continuously decrease due to inevitable spontaneous processes (diffusion, heat conductivity, etc.), but at the same time they restore as a result of self-organization. This means the circulation of counter spontaneous processes (eroding the heterogeneity) and non-spontaneous ones (maintaining it) takes place around the point of bifurcation.	**1. Heterogeneous structure and the counter processes of synthesis and destruction.** Distinct heterogeneity of a cell is displayed on the ionic-molecular level as on the structural level. It is maintained by extraordinary complex circulation of the spontaneous processes (decreasing the gradients) and the non-spontaneous ones (increasing them) providing simultaneously synthesis and destruction (assimilation and dissimilation). Thus, spontaneous diffusion decreases concentration gradients, while non-spontaneous active transport increases them at the expense of energy expenditure. Permanent ionic gradients are maintained between a cell and the environment, their reversible changes support electric field gradients. Structural heterogeneity is more obvious in eucariotic cells: the nucleus, protoplasm, mitochondrions, endoplasmatic network, structures of Goldzhi complex are surrounded and permeated by plentiful membranes fixing gradients of concentrations, pressure, and electric field.

Table 1. (Continued).

Main properties of a chemical system at the bifurcate state	The corresponding characteristics of a living cell
2. Continuous waves and re-arrangement of molecules. The particles are organized by fluctuations through cooperative events and synergism. A competition between the fluctuations continuously changes their radii of correlation. It follows that from time to time the particles transit from one temporal association to another. As a result, continuous re-arrangement of the particles takes place.	**2. Continuous dynamic processes and re-arrangement of molecules/atoms.** A cell is an assembly-line system in which manifold dynamic processes proceed uninterruptedly. Permanent chemical nonequilibrium between a cell and the outside world supports continuity of the processes. Re-arrangement of molecules takes place as due to their directed transfer as through their transformations during nucleo-protein interaction, ATP synthesis and decomposition, etc.
3. Integrity through cooperative processes. At the critical point the dominant fluctuation integrates all particles because its radius of correlation expands to the entire system. The particles are perceptive each other on macroscopic distances due to long-range links and the cooperative effect. The echo of any local event spreads over the whole system.	**3. Integrity based on cooperative interaction.** Integrity of a cell is maintained by the organization and regulation of cellular functions on different hierarchical levels. The organization of an entire cell appears in its organic process of division, coordinate movement, etc. The local structures integrate own parts and molecules as well, being simultaneously dirigible subsystems of the cell. Thus, mitochondrions may contain DNA and be able to self-reproduce, but their reproductive division is dependent on the nucleus.
4. Uninterrupted exchange by matter and energy with the surroundings (open system). Powerful fluctuations arising around the critical point open the system and maintain the exchange by matter and energy, as well as of simple structural information, with the surroundings. Through the exchange the system acquires extraordinary sensitivity to occurrences in the outside world.	**4. Continuous exchange of a cell of matter, energy and information with the environment (open system).** Existence of any organism is maintained through the various exchange processes with the outside world. An organism is highly sensitive to external changes and responds to them on different hierarchical levels. Thus, ions and molecules move through the plasmatic membrane both inside and outside of a cell. Large molecules and particles can be actively extracted from the environment by means of the specific membranes' properties. Unicellular organisms may respond by using complex taxis.

This thesis leads to the paradoxical situation: the bifurcate state should be considered as a starting-point for living systems, but the bifurcation point is in principle unstable. A system cannot be at the bifurcation point for a long time, because the occurrence of many accidental changes inevitably turns its development either into the initial state or into the advanced one. However, nature on Earth found a way to prolong a bifurcate state of the prebiotic microsystems about four billion years ago. The permissible opportunities for this to occur are investigated below.

BISTATE (PULSING) PREBIOTIC MICROSYSTEMS

The period of a bifurcate transition is characterized by the appearance of two opposite forces in a system. They simultaneously strive to turn development of a system into the opposed initial and advanced states. Action of these forces dualizes the system's structure, maintains the internal tension, and keeps the bifurcate properties considered above. The bifurcation point designates peak of the bifurcate period. To prevent of the completed transition to the initial or advanced stable states that would lead to loss of the critical properties, an organic microsystem must oscillate around the bifurcation point staying inside the bifurcate area all the time. Such oscillation in the microsystem might be supported by the optimum oscillations in the surroundings, because any system at the critical state is open and very sensitive to the external changes.

The type of a system that continuously oscillates around the bifurcation point between two polar stable states is called a *bi-state system,* or, in short, a *bisystem* (Kompanichenko, 2004, 2008). In fact, a system of the type is pulsing. The biforked in this manner system simultaneously exists in two autonomous and interrelated states, which become dominating in turn due to oscillations around the central bifurcation point. A bistate organic microsystem may appear under oscillating conditions in the maternal medium, when the regular inversions of the responsible parameters maintain the relative equilibrium between the competing nonequilibrium states. Once formed, the bistate system resists its own destruction through adequate response to external actions. This opportunity follows the Le Chatelier principle: any action to a chemical system due to change of the external conditions initiates the counteraction.

The principal scheme of a bistate system is shown on Figure 2. The point of instability repulses the co-structures to the opposite sides from the axis of smooth subsymmetry (Figure 2 (a)). A bisystem simultaneously strives to evolve in accordance with three main potential trends peculiar to a system during the bifurcate transition: to the conservative state (Trend C), to the advanced higher-organized state (A) and to the advanced lower-organized (B) state (Figure 1). Maintenance of the relative equilibrium between the competing co-structures works in the direction to their balanced separation that is finally displayed in the organic division of a bisystem into two sub-identical components (Figure 2 (b)).

According to the systemic conception, first cells derived of the bistate prebiotic microsystems and converted their method of organization. This thesis can be confirmed by the obvious correlation between the theoretically distinguished properties of a bistate prebiotic system and the real characteristics of a living organism (Table 2). The table shows that the intermediate position of a bistate system between "life and non-life" allows in principle to explain the fission of a cell, inner biophysical and biochemical rhythms, motion of unicellular organisms by means of oscillations of flagellums, accidental mutations in the genetic structures, morphological subsymmetry (dissymmetry) of living beings, their ability to correct own behaviour and modify functions, etc. The theoretically substantiated process of the dichotomy of a bistate system correlates with the real methods of dichotomy of simple and complex cells - amitosis, mitosis, and meiosis (Figure 3).

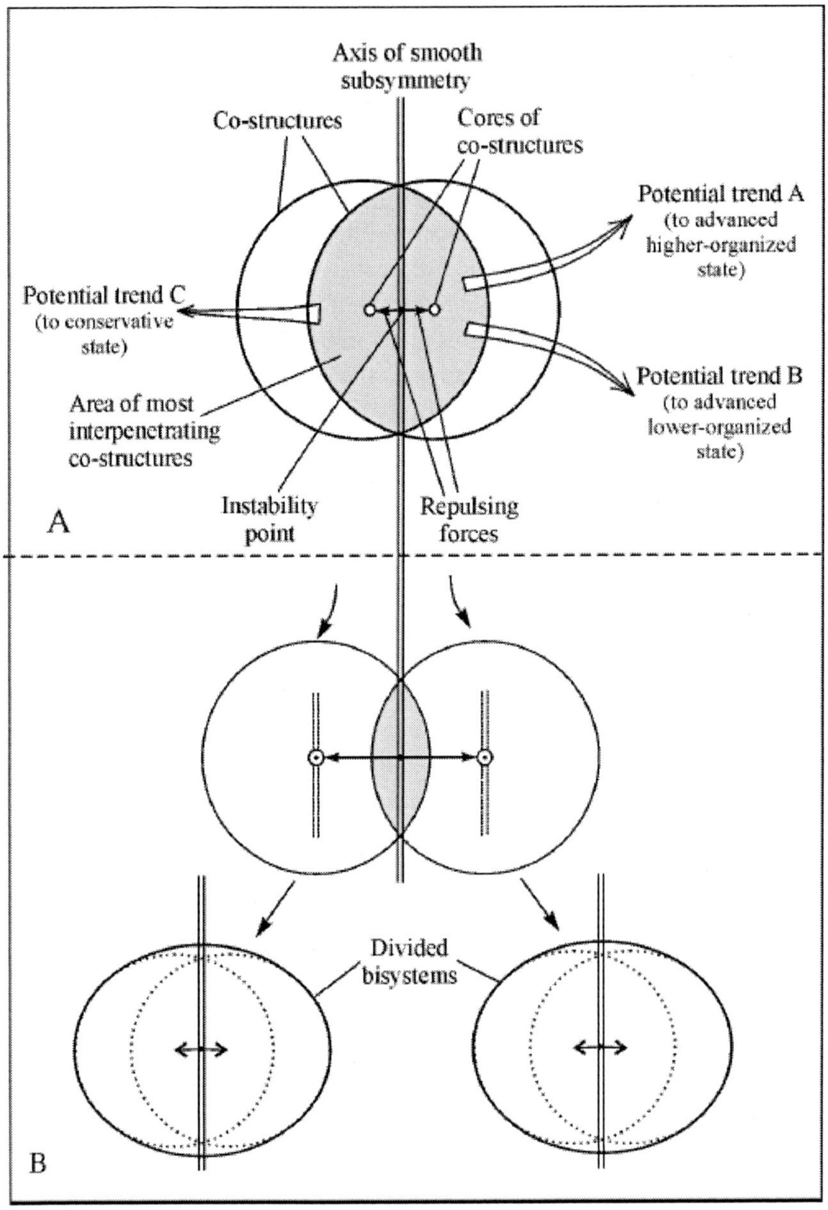

Figure 2. Principal scheme of a bistate system (A) and its dichotomy (B).

Table 2. Correlation between the theoretically based properties of a bistate system and the corresponding real characteristics of a living organism

Main properties of a bistate system	The corresponding characteristics of a living organism
1. Latent biforked structure consisting of two autonomous co-structures divided by the plane of smooth subsymmetry. The plane is maintained by relative equilibrium between the opposite forces. Because the co-structures develop in different directions they are not absolutely identical. That means a bistate system is characterized with the paradoxical combination of the tendency to symmetry and asymmetry that results in its specific dissymmetric method of organization.	1. The biforked structure of a simple procariotic or eucariotic cell is latent and displays during the final fission into two sub-identical units. The biforked structure of advanced multicellular organisms is evident: they possess pair limbs, eyes, ears, cerebral hemispheres, etc., which are developed into different sides from the coincident planes of smooth subsymmetry. Disymmetry is a peculiar property of living organisms.
2. Dichotomy at the end of normal cycle of the existence. This property follows from the paradoxical combination of instability and integrity in a bistate system. Maintenance of the relative balance on a background of the repulsive forces should inevitably lead to the final division of the prebiotic bisystem into two sub-identical components. Figuratively speaking, normal cycle of its existence represents the gradual 'sawing' combined with the tendency to form sub-identical dichotomous units.	2. Fission of a cell at the end of normal cycle of the existence. The division begins with self-replication and divergence of the chromosomes in the core of a cell.
3. Oscillating character of existence related to periodic change of the dominant states. Oscillations of all inner processes support stability of a bisystem and sustain the internal tension that is a source of incessant rearrangement of molecules. Ceasing of the oscillations constrains a bisystem to leave the bifurcate area with a following irreversible transition into one of the stable equilibrium states.	3. Various inner processes in a living organism have oscillating character, for instance, oscillations of flagellums in independent eucariotic cells or hierarchical biophysical rhythms in advanced forms of life.
4. Display of mutability and heredity maintaining the ability to self-evolve. According to the theory of dissipative structures, each forward transition over the bifurcation point (from the initial into advanced state) brings some accidental changes in the chemical system. During the back transition a system strives to return into the initial state that is altered already because of the changes that have occurred. So, a reverse transition can be considered as a way towards conservation of the previous state of a chemical system. In this context, regular oscillations around the bifurcation point occurring in a bistate chemical system work step by step towards accumulation of the accidental changes that provides its ability to evolve. The forward transitions initiate new transformations in a bisystem (mutability), while the reverse ones retain its state (heredity).	4. Mutability and heredity are well-known processes investigated first of all in the framework of evolutional biology and genetics. Periodic mutations reorganize genetic structures at different hierarchical levels (in a gene, chromosome and entire genome).

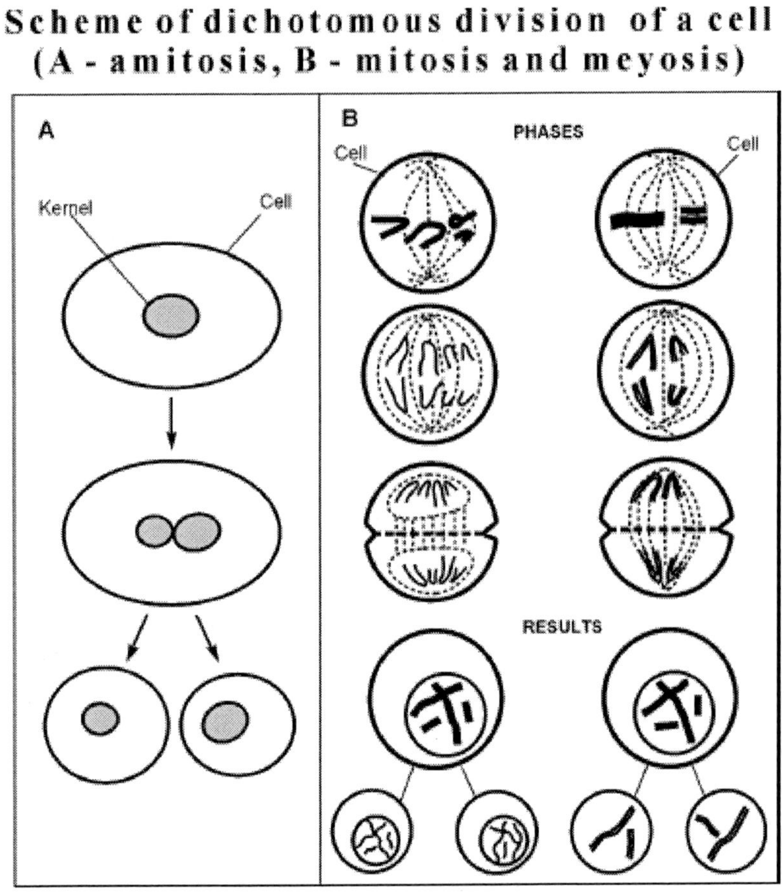

Figure 3. Scheme of dichotomy of a cell (A – amitosis, B – mitosis and meiosis).

EMERGENCE O LIVING PROBIONTS AND THEIR COMMUNITIES

As it was based, a prebiotic bistate microsystem possesses many properties that are essential for life. However, it is still not characterized with the following key unique properties of biological systems listed in the introduction. The essence of the key biological properties could be boiled down to the following thesis: a viable organism concentrates free energy (F) and information (I) through the ability to intensify and make expedient reaction to external changes. Rewording, it is able to keep the positive F and I gradients with respect to the environment. By means of the positive free energy gradient a viable living system is able to transform the environment more efficiently than the environment transforms it (this demonstrates the ability for intensified counteraction). Due to the positive information gradient, a viable biological system knows about the environment more than the environment knows about it (this is a basis for its expedient behaviour). To keep this path of existence, the inflow of free energy F (providing the ability to do a work) into a biological system must be higher than inflow of entropy S (devaluing the ability to carry out a work). The entropy contribution is connected with the total input of the spontaneous (basic) processes in a natural system, while the free energy contribution is related with the total input of non-spontaneous (coupled) processes. That means the ratio 'free energy contribution to entropy contribution' can be used to fix the approximate balance between these counter universal processes. The universal basic and coupled processes embrace all processes that proceed in a system. All non-living natural systems are characterized with negative F/S ratio, while viable living systems possess positive value of the ratio. The permissible method of the inversion of the ratio F/S contribution is theoretically outlined below.

Being a non-living system, a prebiotic bistate microsystem is characterized by a negative total F/S ratio. However, because of the oscillating character of existence the F/S ratio in such a system changes all the time. Besides, sometimes amplitude of the swing of the ratio may rise extraordinary due to a radical constructive or destructive transformation initiated by the accumulation of accidental changes. In the case high amplitude, from time to time the free energy contribution exceeds the entropy contribution although the average long-term F/S balance is still negative (Figure 4-I).

In the changeable conditions prebiotic bistate microsystems undergo periodic stress. The stress affecting a bistate prebiotic microsystem for the period of negative F/S balance leads to its degradation because of the prevailing entropy contribution (Figure 4-III). However, in the case of the period of positive F/S ratio a stress stimulates intensive reorganization of the microsystem possessing an excess of free energy. The surplus of free energy can be used for constructive transformations in the bistate microsystem as a respond to stress, with the following rise of free energy production due to more efficient organization of the network of chemical reactions. The enhanced inflow of free energy (as well as information) in the microsystem increases the positive F/S balance further and facilitates new constructive transformations. Several successive actions from the outside world having an influence for the favorable period may result in an increasing "avalanche-like" constructive conversion of the prebiotic bisystem which turn the *average F/S balance* onto the positive branch (Figure 4-II). In this way probionts on the early Earth acquired the *first unique biological property* – the ability to concentrate free energy and information. The process of free energy concentration was maintained through the second *unique biological property* - the ability to exhibit an intensified counteraction to external influences. Like a free energy, information concentrated in a probiont faster than decomposed. This inverted direction of the information exchange gave rise the formation of *bioinformation*. New quality of information of this kind consisted in the appearance of the functional sequence complexity and programming (Abel, Trevors, 2005). The primary bioinformation that was involved into the network of biochemical reactions allowed a probiont to execute expedient counteraction to external influences (the *third unique biological property)*. The incessant self-renovating processes in a viable probiont combined with its tendency to dichotomy (that appeared at the previous stage - formation of the prebiotic bistate microsystem) represent the *forth unique biological property* – regular self-renovation at different levels (from the continuous molecular reorganization up to selfreproduction).

The inversion of the F/S ratio leads to remarkable changes in the network of chemical reactions. This radical transformation gives rise to metabolism of an initial biological organism and transition of chemical reactions from *in vitro* into *in vivo*. Once formed, a probiont has opportunities to support existence because of the Le Chatelier principle: under regular external actions a probiont persistently develops a counteraction directed to maintain its own structural and functional stability, i.e. the biological method of organization (Kompanichenko, 2008a).

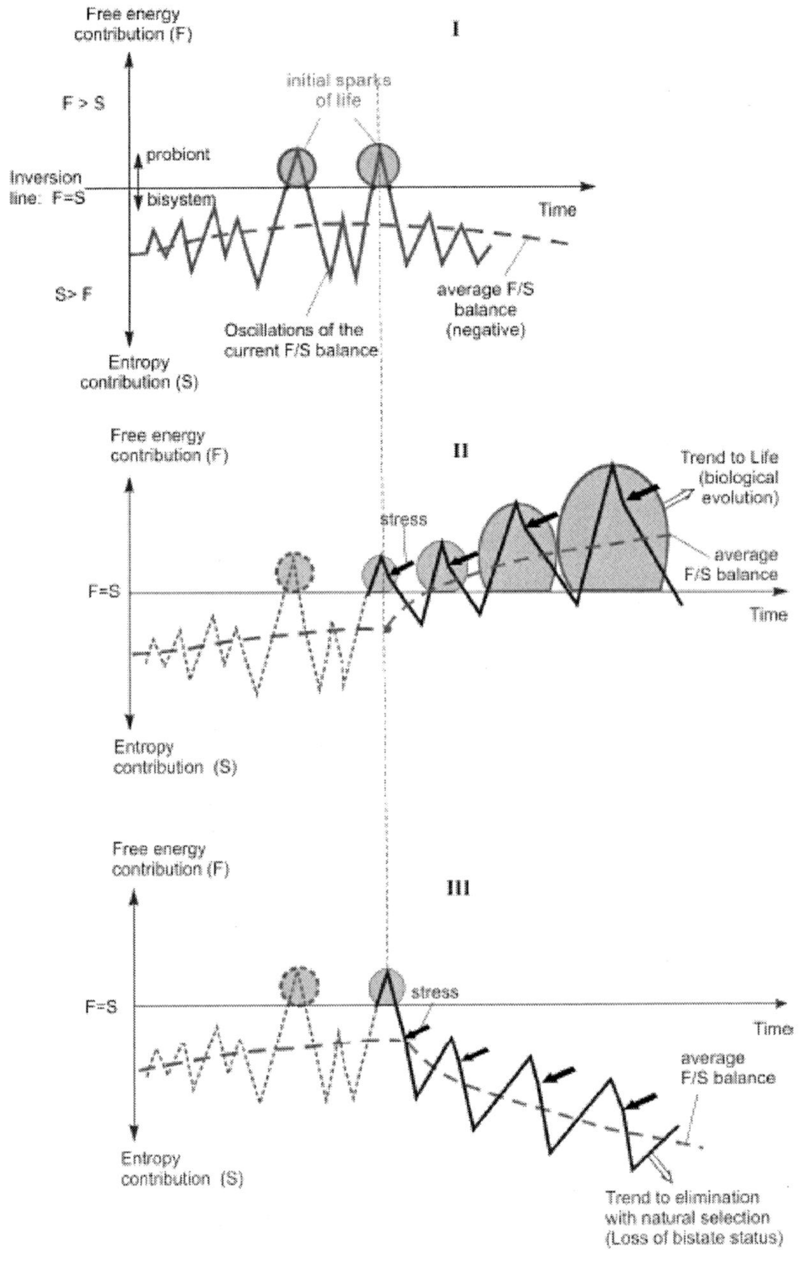

Figure 4. Oscillations of the balance 'Free energy contribution/Entropy contribution' in a prebiotic bistate (micro)system. The different trends of change of the balance: (I) Initial sparks of life in the bistate microsystem that appear because of the oscillating character of its existence; (II) Transformation of a bistate prebiotic microsystem into a primary living unit (probiont) through efficient responses to changes in the outside world (stresses) executed at the periods of the positive F/S balance; (III) Degradation of the bistate microsystem due to the changes influenced at the periods of the negative F/S balance.

The proposed mechanism of probionts formation is statistical: only part of prebiotic bistate microsystems can be transformed into probionts while the rest ones should loss their bistate status due to natural selection at this prebiotic stage. So, the statistically sufficient quantity of the bistate microsystems is probably necessary to produce small number of probionts. The mechanism itself is based on the synergy effect as a capital factor of the maintenance of a system. On this basis the following thesis can be formulated: just a community of probionts, not a single probiont, should be considered as a minimum selfsufficient unit of the primary life.

Optimum fluctuations of the parameters in the geological cradle of life were necessary as to keep stability of prebiotic bistate microsystems (oscillating around the point of bifurcation) as to realize the ability of probionts to exhibit the intensified and expedient reaction to external changes. Such conditions are most peculiar to hydrothermal medium where stable regimes of existence interchange with powerful fluctuations that are generated by tectonicmagmatic activity of our planet. In this context, initial communities of probionts probably appeared around the hot springs in the shallow or deep zones of the Hadean ocean. Besides, the primary communities might originate around discharges of terrestrial hydrothermal systems. These communities, which probably were initially isolated of each other, respond to the embryonic stage of the evolution of ecological systems.

Biological Evolution from Probionts to Progenotes

The progenote that was theoretically reconstructed by C. Woese (1987) can be considered as an intermediate form of life between probionts and prokaryotes. It is correlated with the Last Common Universal Ancestor (LUCA) in a structure of the extinct part of the Phylogenetic Tree. Organisms of this type could have had a genotype and phenotype (i.e., information stored in a quiescent [replicative] form in one class of molecules that was also manifested in an active [functional] form in another), but their genes would for the most part have been physically separate units; they would not be organized into large contiguous linear arrays. Proteins would have been small or nonunique sequence or both. It follows of the reason that proteins of normal size could not have been synthesized without introducing (many) errors. As a consequence, enzymes would not be as accurate and specific as their modern counterparts. The progenote reasonably had error rates two or three orders of magnitude greater then found in cells today. To keep existence of minimally functional cells, genes should be disjoint, and they could have existed in high copy numbers. Probably, the informational macromolecule at this stage was RNA, the functional form of nucleic acid today, not DNA.

RNA is a complex macromolecule therefore the initial nucleoprotein interaction had to appear at the earlier stage of the biological evolution – in a probiont. At that period chemical reactions in a probiont were already involved into the organized network of metabolic processes *in Vivo*. The new biological organization facilitated an increase of thermal stability of nucleic acids. High average temperature of the Hadean ocean (70-80°C that could be higher around black smokers) probably didn't allow even short chains of nucleic acids to exist in abiotic bistate microsystems, at the exception of short periods of life sparks in them during the temporal positive F/S ratio. Bulk of the information in prebiotic bisystems was probably contained in the autocatalytic network (existence of a system possessed such level of information retention is supposed by C.Woese too). New opportunities for the synthesis of polynucleotide chains appeared immediately a prebiotic bisystem had transformed into probiont. Thermal stability of amino acids is higher than nucleic acids. In a general case, the stability of amino acids raises when the degree of their polymerization grows. Therefore, a probiont, inheriting from the previous bisystem oscillating character of existence and incessant circulation of matter/energy, and acquiring the reverse direction of the inner processes (F>S), should inevitably involve into the process of own "self-building" differently organized chains of amino acids and nucleic acids, which functionally complemented each other. Lipids (that play an important role in high temperature transformations of organic matter) and some other compounds should be involved in this process as well.

Biological Evolution from Progenotes to Prokaryotes

Evolution of progenotes into prokaryotic cells proceeded through significant rise of a level of the translation accuracy, elongation of strings of the both classes of biopolymers, the formation of DNA and a genome (Woese, 1987). This process led to an extraordinary raise of the structural-functional complexity of living beings and their communities. In the ecosystemic aspect, this process was characterized with the emergence of new species and appearance of parallel ways in the earliest biological evolution. Existence of manifold cross branches in the Phylogenetic Tree of prokaryotes was emphasized, in particular, by Doolittle (2000). According to the wide-spread opinion among microbial ecologists, (hyper)thermophilic Archaea and Bacteria are at the root of the Phylogenetic Tree (Stetter, 1995; Woese, 1987; Xue et al, 2003), although some of them could evolve from mesophilic forms at later stages of biological evolution. The opinion coincides with the above evaluations of temperature conditions in the hydrosphere of the early Earth. Emergence of species had led to conversion of the initial communities of probionts and progenotes ("embryos" of ecosystems) into the real ecological systems, which were hierarchically organized of different microbial species. Due to their expansion life spread over the entire Earth's surface film. It was a starting point of the microbial biosphere development.

Conclusion

The proposed method of the transition from abiotic organic microsystems to living cells needs further specification and experimental proof. Some relevant kinds of experimental research can be suggested. First, various prebiotic models should be explored at the bifurcate state and under oscillating conditions in the experimental chamber. Exploration of behaviour of organic assemblies under such far-from-equilibrium conditions allows us to concentrate attention just on the moment of the emergence of the key biological properties. Then, it seems necessary to determine the regime of fluctuations (amplitudes, frequencies) of thermodynamic and physico-chemical parameters in the hydrothermal medium that is the most optimal for the transition. The oscillations can be explored in the course of natural experimental research with the following using of the obtained data to conduct laboratory experiments. The investigation of the extinct types of microorganisms at the root of the Phylogenetic Tree should be brought to the probiont's level. Through these and other relevant scientific research the fundamental regularities of the biosphere functioning that were founded at the time of its origin can be clarified. Later on these regularities might serve as a scientific basis that would be helpful to control the global ecological changes on our planet.

REFERENCES

Abel, D. L. & Trevors, J. T. (2005). Three subsets of sequence complexity and their relevance to biopolymeric information. *Theoretical Biology and Medical Modeling, 2*, 29doi:10.1186/1742-4682-2-29

Chaussidon, M. & Franquis, R. (2006). Constrains from silicon isotopic compositions on seawater temperature in the Archaean. Abstracts of 6th European Workshop on Astrobiology, Lyon, October.

Deamer, D. W. (1986). Role of amphiphilic compounds in the evolution of membrane structure on early Earth. *Origins Life, 17*, 3-25.

Deamer, D. W. (2004). Prebiotic amphiphilic compounds. In: J., Seckbach (Eds.), Origins. (75-89). Netherlands, ST: Kluwer.

Doolittle, W. F. (2000). Uprooting the Tree of Life. Sci. Am., February, 90-95.

Ehrenfreund, P., Rasmussen, S., Cleaves, J. & Chen, L. (2006). Experimentally tracing the key steps in the origin of life: the aromatic world. *Astrobiology, 6(3)*, 490-520.

Fox, S. & Dose, K. (1977). Molecular evolution and the origin of life. New York, ST: Dekker,

Fox, S., Balin, P., Pappelis, A. & Yu, B. (1996). Experimental retracement of terrestrial origin of an excitable cell: was it predictable? In: Chela-Flores, J &. Raulin, F (Eds.) Chemical evolution: *Physics of the origin and evolution of life*, (21-32). Netherlands, ST: Kluwer.

Gilbert, W. (1986). The RNA World. *Nature, 319*, 618.

Haken, H. (1978). Synergetics. Berlin, New York, ST: Springer-Verlag.

Haken, H. (2003). Special issue: Nonlinear Phenomena in Complex Systems, *5(4)*.

Holm, N. & Andersson, E. (2005). Hydrothermal simulation experiments as a tool for studies for the origin of life on Earth and other terrestrial planets: a review. *Astrobiology, 5(4)*, 444-460.

Joice, G. F & Orgel, L. E. (1993). Prospects for understanding the origin of the RNA world. In: The RNA World (1-25). New York, ST: Gold Spring Harbor laboratory Press.

Knauth, L. P. (1992). Origin and diagenesis of cherts: an isotopic perspective. In. N. Clanet & S. Chaudhuri (Eds.). Isotopic signatures and sedimentary records, Lecture notes in Earth Sciences, No 43 (123-152). Heidelberg, New York, ST: Springer-Verlag.

Knauth, L. P. & Lowe, D. R. (2003). High Archaean climatic temperature inferred from oxygen isotope geochemistry of cherts in the 3.5 Ga Swaziland Supergroup, South Africa. Geol. Soc. *Amer. Bull., 115(5)*, 566-580.

Kompanichenko, V. N. (2003). Distinctive properties of biological systems: the allround comparison with other natural systems. *Frontier Perspectives, 12(1)*, 23-35.

Kompanichenko, V. N. (2004). Systemic approach to the origin of life. *Frontier Perspectives, 13(1)*, 22-40.

Kompanichenko, V. N. (2008). Three stages of the origin-of-life process: bifurcation, stabilization and inversion. *Int. J. Astrobiology*, *7(1)*, 27-46.

Kompanichenko, V. N. (2009). Changeable hydrothermal media as a potential cradle of life on a planet. *Planetary and Space Science*, *57*, 468-476.

Luisi, P. L. (2000). The relevance of supramolecular chemistry for the origin of life. *Advances in Supramolecular Chemistry*, *6*, 287-307.

Mukhin, L. M., Bondarev, V. B., Kalinichenko, V. I. & Vakin, E. A. (1979). Amino acids in hydrotherms of Southern Kamchatka. *Doklady An Ussr*, *244(4)*, 974-977.

Nicolis, G. & Prigogine, I. (1977). Self-organization in nonequilibrium systems. New York, ST: Wiley.

Prigogine, I. (1997). End of Certainty. The Free Press.

Prigogine, I. & Stengers, I. (1984). Order out of chaos. New York, ST: Bantam.

Russell, M. J. (2003). The importance of being alkaline. *Science*, *302*, 580-581.

Schidlowski, M. (1997). Carbon isotopes and the oldest record of life: potential and limits. SPIE, *3111*, 462-471.

Schopf, J. W., Kudryavtsev, A. B., Argesti, D. G., Wdowiak, T. G. & Czaja, A. D. (2002). Laser-Raman imagery of Earth's earliest fossils. *Nature*, *416*, 73-76.

Schwartzman, D. (1999). Life, temperature, and the Earth. New York, ST: Columbia University Press.

Stetter, K. O. (1995). Microbial life in hyperthermal environments. *ASM News*, *61(6)*, 328-340.

Westall, F., de Vries, T. & Nijaman, W. (2006). The 3.466 Ga "Kitty's gap Chert" an early Archaean microbial ecosystem. In: W. U., Reymond & R. L., Gibson (Eds.) Processes on the early Earth (105-131). *Geological Society of America*, *405*.

Woese, C. R. (1987). Bacterial evolution. *Microbiol. Rev.*, *51(2)*, 221-271.

Xue, H., Tong, K., Marck, C., Grosjean, H. & Wong, J. T. F. (2003). Transfer RNA paralogs: evidence for genetic code-amino acid biosynthesis coevolution and an archaean root of life. *Gene*, *310*, 59-66.

In: Astrobiology: Physical Origin, Biological Evolution... ISBN: 978-1-60741-290-8
Editors: S. Hegedus, J. Csonka, pp. 131-141

Chapter 7

VISUALIZATION IN THE EDUCATION OF ASTROBIOLOGY

Akos Kereszturi*

Collegium Budapest Institute for Advanced Study, Hungarian Astronomical Association, Karoly Nagy Astronomical Foundation

ABSTRACT

This chapter summarizes the experiences regarding the general knowledge of undergraduate students in astrobiology, and presents methods we used in the education of astrobiology at university level in Hungary. Based on these, we found the following important issues. They have good general knowledge in the fields, which are well known for decades. Liquid water is thought to be widespread by the students, but they lack important information (like oceans inside icy moons) in the recent results. The most important is to develop and improve methods for the students to use the knowledge learned at different subjects more frequent in astrobiology. Useful examples, like the astrobiology- and Drake-matrixes are presented to compare different environments based on numerical values, graphics and tables.

INTRODUCTION

This chapter summarizes our results and experiences on the development of astrobiology related educational modules and their integration into the teaching of natural sciences at university level. Our aim was to survey the students' basic knowledge on astrobiology, and develop materials, which help the undergraduate education.

Here we do not give detailed description of all the education aspects, but present small and practical ideas together with examples, which were useful in our course in the last years [Kereszturi 2004]. These materials were developed under the cooperation between Eotvos

* Corresponding author: E-mail akos@colbud.hu

Lorand University of Sciences, the Polaris Observatory maintained by the Hungarian Astronomical Association and the Karoly Nagy Astronomical Foundation.

There are several works on the educational aspects of astrobiology, analyzing its multidisciplinary aspect [Bhathal 2003, Methot and Finley 2002, National Research Council 2005], simple, essential questions [Slater 2005], and teaching modules [Kolb 2005], partly in secondary and higher education too [Rodrigues 2005, Foster and Drew 2009], and even for the public [Race 2002]. Several teaching materials are available today, like classical textbooks [Morrison 2005], educational games [Rodrigues and Carrapiço 2004], materials emphasize the latest result [Linch 2002], also ideas for the use of online methods [Edward 2002; Madison 2003, Simon and Kereszturi 2009] as well as hands-on experiments [Fauerbach et al. 2007]. Astrobiology is a popular topic in the education [Morrison 2004], but challenging at the same time, because it covers several fields of different natural sciences.

We would like to give information for the readers by focusing to on educational modules and methods, which help fuse knowledge from various fields of natural sciences. Such synthesis is of high importance at university level, to understand connections and trigger interdisciplinary research ideas [Luen 2006] by the students in their own field of specialization in the future.

METHODS

We surveyed the students' knowledge during the course with tests contain 1. simple estimation of numerical parameters or relative ordering, and 2. complex comparison to connect different environments and/or processes. The results of the tests were analyzed by statistical methods.

We aimed at to fuse two main aspects during the course: the presentation of *1. up-to date scientific information* with proven background knowledge, and the application of *2. creative and new ideas* at poorly known fields. Regarding the first issue we used numerical parameters (p, T, pH etc.) as well as published and accepted theories. Regarding the second issue, where unknown or poorly known process was hypothesized, we used hypothetical range of values, which are considered to be realistic based on our current knowledge, and also outlined simplified version of suspected process/environments, to show causal connections. We tried to trigger new ideas from the students, where the important parameters are not well known.

Here we summarize our results in two groups: 1. Visualization and comparison, 2. Water and planetary evolution [Kereszturi 2004]. In every topic we present the result of a representative survey that gave insight into the students' general knowledge, and after it a useful method or example is presented that helps to visualize and understand the missing issues.

VISUALIZATION AND COMPARISON

At the beginning of the course it is worth giving a rough overview of astrobiology as a whole scientific field. The aim was to show the connections between different disciplines, and how these topics point into the same direction: toward the understanding of the possibility of life beyond the Earth.

Table 1 shows the result of a survey to find out which topics are causally connected, based on the students general knowledge in astronomy, physics and earth sciences. The first column lists different environments, and the other columns show the formation, decomposition of chemical elements and organic molecules, and the presence of liquid water. The students had to put X in those cells, where in that environment (different rows) certain processes or liquid water is present. In the cells the percentage of the students is indicated, how many of them think connection exists between certain environments and processes / phases.

Table 1 gives insight where the students think connections or interesting relations are present. Regarding the formation of elements they are mostly wrong, and imagine nucleosynthesis is present in far more environments than in reality – and/or misinterpret the exact meaning of the word "element". The case for molecule formation and decomposition is better, but they think liquid water is present in more environment types than in reality. The result suggests they are ready to connect different fields, but the background knowledge is not accurate.

To show connections in astrobiology, we completed a matrix-shaped table, and named it astrobiology matrix (Figure 1) to visualize the different fields and their relations in astrobiology [Mizser and Kereszturi 2003]. The left one of the two main vertical columns represent the Earth-science related topics (planetology column), and the right one topics of astronomy (astrophysics column). Along the horizontal and vertical axes various fields of classical sciences are visible.

Table 1. Percentage of students (numbers in cells) who think in different environments (rows) certain processes / phases (columns) are present

	formation of chemical elements	formation of organic molecules	decomposition of molecules	presence of liquid water
main-sequence stars' atmosphere	43	0	7	0
interstellar material	29	50	21	0
growing protoplanets in proplyds	21	21	14	29
collision of planetary boodies	21	29	43	29
volcanic centers on rocky planets	21	36	14	7
surf. of minor planets and comets	0	21	29	28
planetary atmospheres	0	36	14	50
coma of comets	14	7	36	14

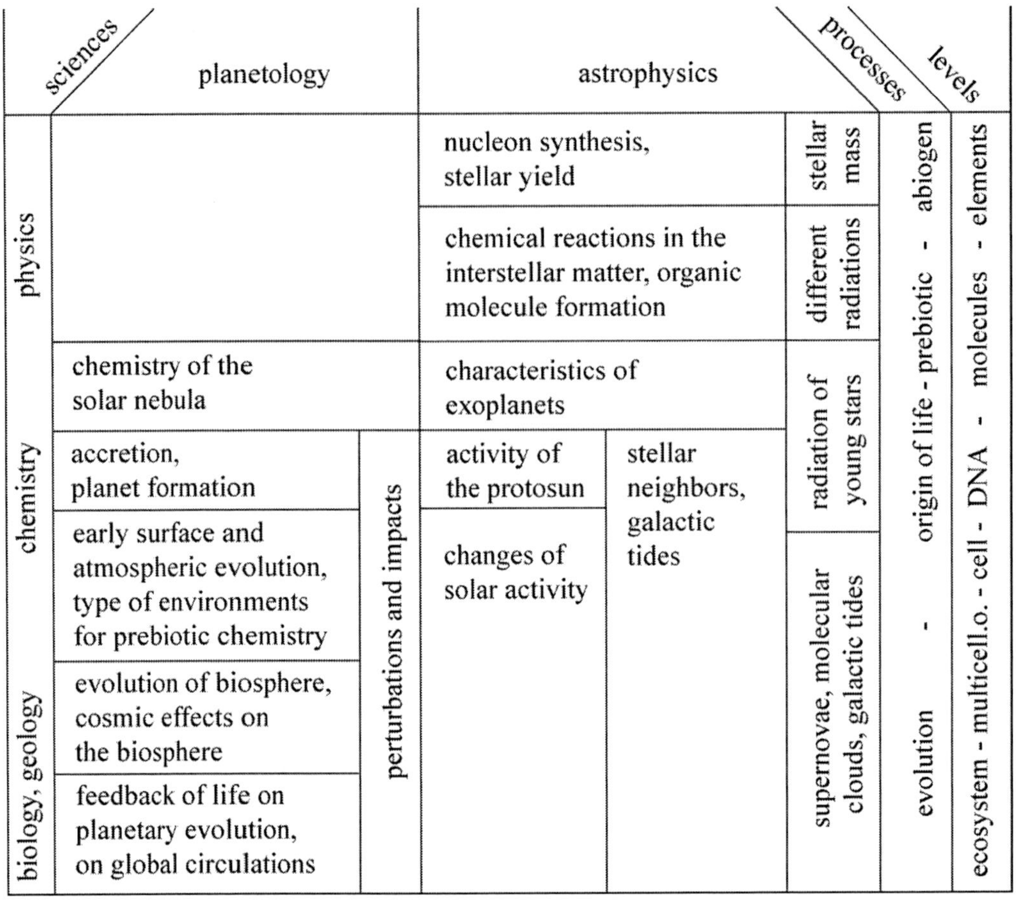

Figure 1. The astrobiology matrix, representing interdisciplinary connections in astrobiology

The narrow topics of research, from the nucleosynthesis and stellar yield, through the chemical evolution of interstellar material and planet formation, toward solar and extrasolar effects on climate and planetary evolution etc. are in the middle of the table. Vertically two "evolutionary" trends are visible from top toward bottom: 1. time passes by toward the bottom, and 2. more complex systems are formed (atoms, molecules, cells, multicellular systems and ecosystems) together with more and more complex processes (abiotic, prebiotic, and biotic phases, and evolution of whole ecosystems).

A survey was made to compare different environments important in astrobiology (Table 2.). In this questioner one had to indicate where organic material and amino-acid are present. (The term "presence" was used for the condition, where a simple chemical analyzer, resembling to those used in the laboratory during the education at university, would find that material.) Based on the survey, some students thought organic material is present in many planets or meteorites, most of them thought is present in interstellar space. Few students thought that amino acids are present anywhere beyond Earth, Mars and other very unique planets, and only percent thought they are inside meteorites.

Table 2. The result of the survey, on the occurrence of organic molecules and amino acids. The cells show the percentage of students think organic material or amino-acids are present at different environments (rows)

environment type	organic material	amino acids
stellar atmospheres	6	0
interstellar material	44	7
average meteorites	11	1
most planets	5	0
only unique planets (Mars, Earth)	33	56
nearly nowhere, except Earth	0	36

Table 3. Average estimated and measured temperature values in °C for Earth and Mars

Location /period	estimated	Measured
Earth global average	15	14
Mars global average	-17	-63
Mars equator summer noon	20	7
Mars equator summer midnight	-30	-70
Earth southern pole summer noon	-12	-10
Earth south pole winter midnight	-52	-70
Mars CO_2 cap winter midnight	-95	-120
Mars CO_2 cap summer noon	-30	-120
Mars water ice cap summer noon	-35	-73

To compare the astrobiological potential of different environments, we made a table, based on the Drake-formulae, and named Drake matrix. Frank Drake published the first version of its formula in 1961 and its versions summarize the components regarding to the possibility of life and intelligence beyond the Earth in our Galaxy. In the last years there were important discoveries in various fields of astrobiology. The Drake formulae still summarize our up to date knowledge a simple and elegant way, but it is useful to extend it on other ways.

The Drake-matrix shows the basic requirements, which are necessary for the origin of life like on Earth: chemical conditions (metal content, molecular stability, liquid water), energy source (radioactive, tidal, electromagnetic radiation) and duration of the favorable conditions. It is difficult to give exact values for these parameters, only a theoretical approach is summarized in the table. The Drake matrix (Figure 2.) gives possibilities to explain the reasons in the differences of astrobiology potential for various celestial bodies. It also contains such bodies, which probably don't hold favorable conditions for life, but represent important steps in the evolution of material in the Universe, and give possibilities for comparison.

To estimate the astrobilogical potential of a planetary body, or a part of its surface, atmosphere and interior, basic physical parameters like pressure, temperature, pH etc. are good indicators. It is useless to memorize many of these parameters, but students must be able to roughly estimate some of them. To find out, which are the well and less accurately known parameters, tests were written to estimate the conditions on different planets and various features on them. One important example is visible in Table 3. on the temperature values estimated by the students and measured in reality, for both Earth and Mars.

It is visible, that students are more familiar with temperature values on Earth than on Mars, and they often overestimate the Martian ones. It also shows an example for the interesting case, that despite students learned important values at other disciplines (like frost point of CO_2) it is not natural for all of them to use here the knowledge learnt in other disciplines.

Great number of possibilities is there how to compare various locations on different planetary bodies. One important guideline is water, and different environments can be marked on the H_2O phase diagram. In the following we show an example to compare the astrobiological potential of local conditions with possible p/T characteristics, to find out where liquid water and heat source exist together. In Figure 3. the average numerical values of pressure and temperature are indicated for impact craters with diameter of several 10 km, geothermal heat sources plus ice as ice chimneys [Hofmann and Kyle 2003] and black smokers – all of these for Earth, Mars, Europa and Titan. The diffuse black streaks represent the range of possible values for these environment types. The arrows show the direction where the real values shift in the case of cooling craters as time passes by.

The comparisons of different environments give possibility for the students to understand what kind of chemical reactions may take place under different conditions, and where are possibilities for complex organic molecule formation, or decomposition. Because the real chemical reactions in any environment depend on many factors, the comparison should be simple. As exact numerical values are not available in every case, we used rough estimations and ranges of possible values during the course.

Drake formulae $N_c = N^* f_p n_e f_l f_i f_c L_c / L_*$

object type	location	materials				energy sources	duration
		metals	organic molecules	liquid water	molecular stability		
interstellar dust	grain surface, interior	moderate	low	no	variable	UV, cosmic rays	variable
asteroids	subsurface	high	high	early	relatively good	UV, cosmic rays, impacts	early, short
icy bodies (comets)	subsurface	moderate	moderate	early	relatively good	UV, cosmic rays, impacts	early, short
icy moons	surface, subsurface	relatively high	moderate	early, tidal	good close to the suface	differentiation, radioactivity, insolation, tides	early, periodic
rocky planets	surface, subsurface	high	high	variable, on surf., subsurf.	good close to the suface	differentiation, radioactivity, insolation, tides	long, decreasing
giant planets	atmosphere	moderate	moderate	atmospheric vapor	good in the atmosphere	contraction	long, decreasing
brown dwarfs	atmosphere	moderate	moderate	late atmos-pheric vapor	good in the late atmosphere	contraction	late, long, decreasing

Figure 2. The Drake matrix. The components of the Drake-formulae are visible at the top, and related ratios or relative weights are in the columns. Different rows are for different environments.

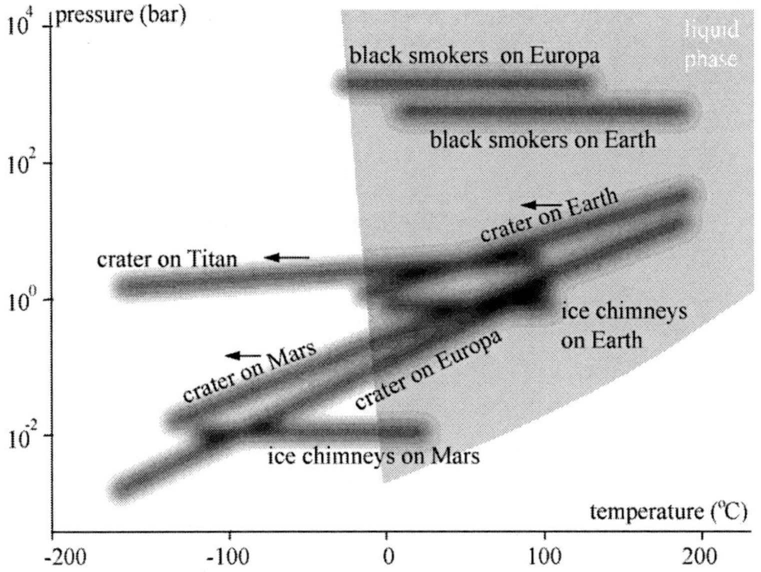

Figure 3. Heat sources on Earth, Mars, Europa and Titan on the p/T diagram of H_2O

WATER AND PLANETARY EVOLUTION

Water is essential for the life we know on Earth. Its stable existence points to not only certain pressure and temperature conditions (depending on salt content), but also to the presence of elements heavier than hydrogen and oxygen. As a result, water can serve as good indicator of environments interesting in astrobiology.

We surveyed the students' knowledge on the occurrence of bulk liquid water in the solar system with a simple exercise (Table 4.). Six planetary bodies from the solar system (Callisto, Mars, Europa, Titan, Earth, Venus) were listed, and one had to put them into decreasing order according to their liquid water content (pure and salty water together), based on our current knowledge.

Table 4. Result of the survey on the distribution of bulk liquid water in the Solar System. The students had to arrange these six planetary bodies to ascending order, according to their decreasing absolute water content. The average estimated order values are visible in the middle, and the real values on the right

planetary body	average order	real order
Earth	1.7	3
Europa	2.6	2
Titan	3.3	1
Mars	4.0	5
Callisto	4.0	4
Venus	5.0	6

Most students put Earth to the first place, followed by icy moons with suspected internal oceans. Mars has the same place as Callisto, while Venus is the last for most students. It is obvious that the students have problems with above all in the icy moons' water content.

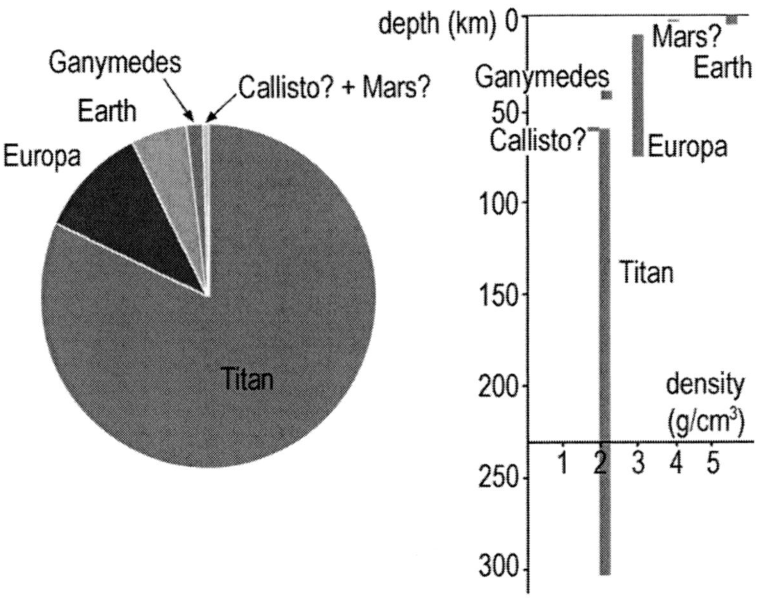

Figure 4. Distribution of water according to its volume on the surface, in the subsurface oceans and pore voids (left) for different planetary bodies, and the proposed depth of these oceans versus the average density of planetary bodies (right)

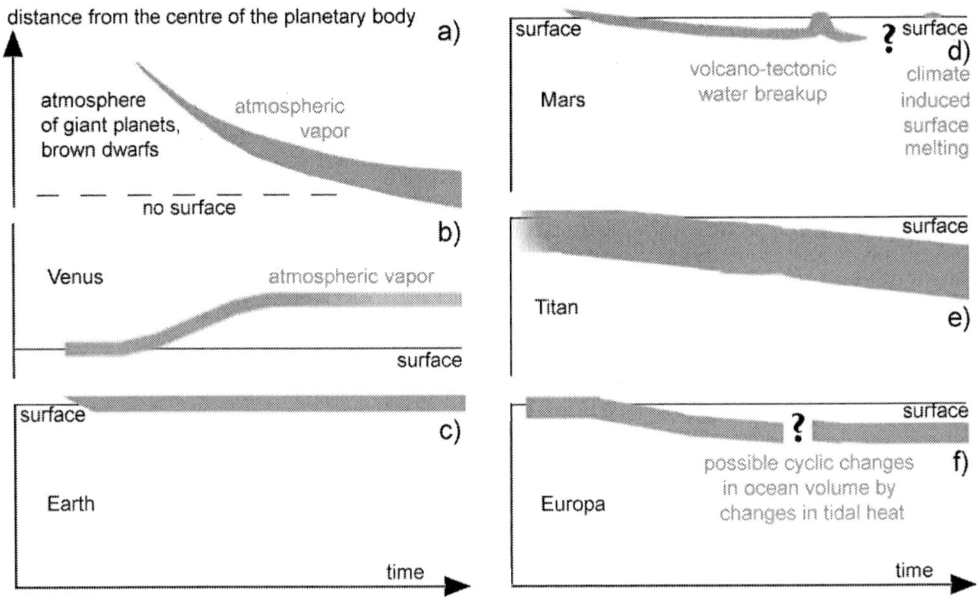

Figure 5. Possible evolutionary tracks of wet environments in various planetary bodies in the Solar System and brown dwarfs

In Figure 4. the volumetric distribution of the hypothesized liquid water is visible in the Solar System, based on current theories and direct as well as indirect observations. We excluded the atmospheric vapor from this figure (the atmospheric water content of the giant planets). On the right-hand side of the figure the depth of the water bodies are indicated versus the average density of the planetary body. Two main groups are represented: the rocky planets with surface or near surface water, and icy moons with deep subsurface oceans.

Water is an important guideline that helps to organize an evolutionary trend for various environments in the point of view of their conditions important for astrobiology. Figure 5. is to compare the changes of wet environments on and inside different planetary bodies. Time passes by horizontally from left to right, the vertical axis represents the height above, or depth below the surface (except where no solid surface is present).

The curves show two main theoretical trends. In the case of cooling environments a), d), e), f), the wet region sinks during planetary evolution. The case of the Earth: subset c) shows nearly continuous presence of the liquid water on the surface, while in the case of a warming planetary body, like Venus in b), the location of wet environment rises as time passes by, and it is in the acidic clouds today.

CONCLUSION

Various methods were tested and improved in the education of astrobiology at university level in Hungary. Based on our experiences, there is a great variety in the knowledge of students: in certain "classical" areas their knowledge is close to those accepted by the scientific community today - but there are weak and missing fields too. Here the information from outside the regular university courses is important, but not enough.

The surveys showed that the "image" of the students in important fields of astrobiology, based on general background knowledge in natural sciences is weak, without specialized education. They have problem to imagine the locations of the formation of different chemical elements, they think elements are able to form in more type of environments than in reality. At the same time they are familiar with the concept that organic material can form and decompose at different environment types. Although nearly no one thinks amino acids could be present at any other locations beside those, which resemble to Earth and Mars – opposite to the well know cases of chondrites.

The students usually overestimate the temperature values of Mars, and despite they learned related issues in other disciplines, it is not always natural them to use those in astrobiology, above all the CO_2 frost point on Mars. Liquid water is also thought to be widespread by the students, but most of them lack the information on liquid water oceans inside icy moons of giant planets.

To compare environments with one or two parameters, real numeric values and their graphical visualization, on a phase diagram of H_2O is useful – while to compare environments based on several parameters together, relative ordering is the effective (like the Drake matrix). To visualize the connections between different fields of astrobiology, a matrix shaped table (astrobiology matrix) was useful where certain topics were indicated together with their relation to classical scientific fields like physics, chemistry etc.

We found the upper mentioned methods useful above all at the beginning of astrobiology course. At an advanced level, when the students are doing their own research but they are still undergraduate [Csorba 2008, Kuti 2009], they are specialized into smaller fields and the education focuses on the use of certain software, model computation or to realize laboratory measurements.

ACKNOWLEDGMENT

This work was sponsored by the Pro Renovanda Hungariae Foundation Science in the Education project, and the Hungarian Astroniomical Association. The cooperation between the Eotvos Lorand University, the Polaris Observatiory, and the Karoly Nagy Astronomical Foundation in coordinating the educational project is also acknowledged.

REFERENCES

Bhathal, R. (2003). Effective Teaching and Learning of Astronomy, *25th meeting of the IAU*, Special Session 4.

Carrapico, F. J., Lourenco, A., Fernandes, L. & Rodrigues, T. (2002). Journey to the origins: the astrobiology paradigm in education, *In Instruments, Methods, and Missions for Astrobiology IV. Proc. SPIE* 4495, 295-300.

Csorba, A. (2008). "A Hellas-medence (Mars) litológiájának és felszínmorfológiájának kapcsolata", Ifjú Szakemberek Ankétja, (in Hungarian).

Edward, E. & Timothy, F. (2002). An Online Astrobiology Course for Teachers, *Astrobiology, 2*, 215-223.

Fauerbach, M., Mon, M. J. & Ueda, T. (2007). Understanding Astrobiology Through Realistic Laboratory Experiences, *Lunar and Planetary Science Conference XXXVIII*, abstract #1338.

Foster, J. S. & Drew, J. C. (2009). Astrobiology Undergraduate Education: Students' Knowledge and Perceptions of the Field. *Astrobiology*. doi:, 10.1089/ast. 2007.0221.

Hoffman, N. & Kyle, P. R. (2003). The Ice Towers of Mt. Erebus as analogues of biological refuges on Mars, *Sixth International Conference on Mars*, abstract #3105.

Horvai, F. & Kereszturi A. (2009). Geology of Mars: new university course in Hungary, *Lunar and Planetary Science Conference XXXX*. abstrct 1673.

Kereszturi, A. (2004). *Planetary environment comparison in the education of astrobiology*, LPS Conference 1070,

Kolb, V. M. (2005). NASA's planetary protection program as an astrobiology teaching module, *Proc. of the SPIE*, 5906, 150-154.

Kuti A. (2009). *Thermal behavior of Dokka crater and its surroundings in the north polar. region of Mars*. 40th Lunar and Planetary Science Conference #1006.

Linich, M. (2002). Life science education in Australia and America: Linking new knowledge with new opportunities, *Proc. of the Australian-American Fulbright Symposium: "Science Education in Partnership"* 55.1.

Luen, T. B. (2006). Astrobiological Themes for Integrative Undergraduate General Science Education, *Astronomy Education Review*, *4*, 110-114.

Maddison, S. T. (2003). Teaching Astrobiology Online, *Proc. of IAU Symposium #213. Bioastronomy: Life Among the Stars* 542.

Methot, L. & Finley, J. (2002). Origin and Evolution of the UCLA AstroBiology Society, In *Bioastronomy 2002: Life Among the Stars, Proceedings of IAU Symposium #213.*, 559.

Mizser A. & Kereszturi A. (2003). The Astrobiology Matrix and the "Drake Matrix" in Education, *Lunar and Planetary Science Conference XXXV*, abstract 1114.

Morrison, D. (2005). Planetary Science as Presented in Astrobiology Textbooks, *Bulletin of the American Astronomical Society.*, *37*, 635.

Morrison, D. (2004). Astrobiology: Not Just Little Green Men, *Bulletin of the American Astronomical Society.*, *36*, 825.

National Research Council, Committee on the Origins and Evolution of Life, (2005). Integrating Astronomy with the Other Disciplines of Astrobiology, in *The Astrophysical Context of Life*, the National Academies Press, 52-56.

Race, M. (2002). The challenges of educating the public about astrobiology via the mass media, *Proc. of the Australian-American Fulbright Symposium 2002: "Science Education in Partnership*, 46.1

Rodrigues, T. & Carrapico, F. (2005). Teaching astrobiology: a scientific and a cultural imperative, *Proc. of the SPIE*, 5906, 146-149.

Rodrigues, T. & Carrapiço, F. (2004). Educational games: a funny way to teach astrobiology, *Proc. of the Third European Workshop on Exo-Astrobiology ESA SP-545*, 277-278.

Simon, T. & Kereszturi, A. (2009). Online astrobiology course in Hungary, *Lunar and Planetary Science Conference XXXX*, abstract 1048.

Slater, T. F. (2005). Capturing Student Interest in Astrobiology through Dilemmas and Paradoxes, *Bulletin of the American Astronomical Society, 37*, 1262.

In: Astrobiology: Physical Origin, Biological Evolution... ISBN: 978-1-60741-290-8
Editors: S. Hegedus, J. Csonka, pp. 143-192

Chapter 8

THE SEARCH FOR LIFE ON MARS: RAMAN SPECTROSCOPY AND ANALYTICAL ASTROBIOLOGY*

***Howell G. M. Edwards*† and Michael D. Hargreaves**
Centre for Astrobiology and Extremophiles Research, Raman Spectroscopy Group,
Division of Chemical and Forensic Sciences, School of Life Sciences,
University of Bradford, UK

ABSTRACT

The analytical application of Raman spectroscopy to the identification of the survival strategies of terrestrial extremophiles through the characterisation of key biomolecular and biological signatures has provided novel information about the roles of protective chemicals synthesised by the organisms under environmentally stressed conditions. The non-destructive sampling requirement and the ability to discriminate between organic biomolecular spectral signatures and those from the inorganic host matrix in the geological record are essential requirements for the assessment of the survival strategies being employed against low wavelength ultraviolet radiation insolation, desiccation and extremes of temperatures found in terrestrial habitats termed Mars analogue sites (Table 1).

The recent announcement of the inclusion of a miniaturised Raman spectrometer as part of a combined Raman /LIBS instrument for the molecular and elemental analytical characterisation of specimens in the Martian surface and subsurface as a component of the European Space Agency ExoMars life-detection suite has focussed attention on the Raman spectra of key biomarkers of life signatures. To search for indicators of extinct or extant life signatures on Mars serves to highlight the importance of Raman spectroscopic techniques in this astrobiological mission. As a forerunner to the ExoMars mission it is imperative that the Raman spectroscopic data from extremophilic colonies from as wide a

---

* A version of this chapter was also published in Planet Mars Research Focus, edited by Lorenzo A. Costas, Nova Science Publishers. It was submitted for appropriate modifications in an effort to encourage wider dissemination of research.

† Corresponding author: E-Mail: h.g.m.edwards@bradford.ac.uk; Tel : 00 44 (0)1274 233787; Fax : 00 44 (0)1274 235350.

possible range of terrestrial sites of relevance to Mars is undertaken. This chapter will address the type of information that is required and which is accessible from these terrestrial studies to adequately prepare the way for the interpretation of analytical scientific data that will emerge from the Mars exploration programme, commencing with the launch of ExoMars in 2013.

INTRODUCTION

Raman spectroscopy is an analytical technique which provides information about the molecular structures and chemical environments of organic and inorganic molecules and molecular ions [1]. The generation of Raman spectra requires the illumination of target specimens by laser radiation, which can be focussed or non-focussed, and for which the "footprint" can range from about one micron to several hundred microns in cross-sectional diameter depending upon the optical design of the sample illuminator. Raman spectra are observed as two sets of wavelength shifted bands from the Rayleigh line, which occurs at the frequency of the incident laser radiation, the long wavelength set of Raman bands comprise the Stokes Raman spectrum and the low wavelength set that of the anti-Stokes Raman spectrum, which is relatively weaker in intensity compared with the Stokes Raman spectrum [2]. Raman spectra are generally measured as wavenumbers (cm^{-1}) which are shifted from the laser excitation wavenumber, namely that of the Rayleigh line.Since Raman spectra are significantly weaker in intensity compared with the Rayleigh line intensity, it is important that the irradiance of the incident laser being used for the observation of the Raman spectra is high, consistent with the requirement that sample damage does not occur. Because the Raman effect has an inverse dependence upon the fourth power of the laser wavelength, much Raman spectroscopy has been carried out in the visible region of the electromagnetic spectrum using blue and green wavelengths. However, the advent of non-visible wavelength diode lasers operating in the ultraviolet and the near-infrared regions has provided some novel sources of excitation of Raman spectra ; currently, there are reports of Raman spectra being excited at wavelengths of 240 nm in the deep ultraviolet and at 1330 nm in the near infrared region and with all other instrumental factors being equal, power for power, the Raman spectrum excited in the ultraviolet will be some 950 times more intense than that recorded in the near infrared region over this wavelength range.

Despite the advantage of greater intrinsic Raman scattered intensity at low excitation wavelengths, there are several good reasons for the adoption of Raman laser excitation at longer wavelengths, as will emerge from the discussion that follows, not the least of which is the sensitivity of the materials under study to the high laser energies in the ultraviolet and the onset of fluorescence associated with the presence of low energy molecular electronic states that are more likely to be probed using exciting radiation at low wavelengths. Fluorescence emission is several orders of magnitude greater than the corresponding Raman scattering generated by the same laser wavelength and relatively intense, broad-band fluorescence backgrounds can generally swamp the much weaker Raman bands from the same sample. An example of this is provided in figure 1; here, the Raman spectra of a *Caloplaca saxicola* epilithic lichen from Crater Cirque, Northern Victoria Land, Antarctica, are shown. The details of the sample location and habitat are provided in Table 2.

Table 1. Antartic Mars analogue cold desert sites

State of liquid water	Antarctic site	Location (latitude and longitude)	Extra-terrestrial analogue	Mars habitat* Gya)	Putative conditions during existence Of Martian habitat	Reference
Aquatic cyanobacterial communities (saline and brackish)	Mars Oasis	71° 52'S, 68° 15'E	Early Mars	4.2-3.8	P > 5 atm T > 0°C CO2, volcanoes	Horneck (1995) Carr (1996)
Flowing surface water	Canada Stream	77° 37S, 163° 03E	Early Mars		Meteorites	McKnight *et al* (1993)
High UVB exposure	Mars Oasis (Ozone Hole)	71° 52'S, 68° 15'E	Earth orbit (ISS)			
Ice-covered	Microbial (cyanobacteria)		Early Mars· Palaeolakes	3 8-3 1	T < 0°C (-20° C)	Wharton *et al.* (1995)
lakes	mats in Dry Valley lakes		Mars: Gusev Crater		Peak T > 0°C	Grin and Cabrol (1997)
	(Lake Hoare) Photosynthesis at low light. UV-tolerance	77° 38S, 161° 51E	(Map: USGS MC-23 SE I-2119)		Carbonate Dissolved gases	Wharton (1994)
Liquid water	Endolithic microbes,	76° 55S,	Porous Martian	3.1-1.5	Peak T < 0°C	Cabrol and Grin (1995)
in porous rocks	inside Beacon sandstone	160° 55E	lithic niches		P c.100 mb	Wynn-Williams et al. (1999)
Or aquifers	at Battleship Promontory (cyanobacteria)					
No liquid water on surface.	Victoria Valley sand dunes (negligible microbes)	77° 20S, 161° 45E		1.5 - present	P ~ 7.4 mb Low CO2 Vacuum	Cameron (1972)
	Horseshoe Mountain (no endolithic microbes)	77° 34S, 159° 57E				Friedmann *et al.* (1994)
	Palaeolake beds (L. Fryxell)	77° 37'S, 163° 09'E	Martian palaeolakes			Doran *et al.* (1998) Wharton *et al* (1995)
Permafrost.	Taylor Valley	77° 44S, 162° 10E				Vorobyova *et al.* (1997)
Geothermal water?	Mount Melbourne (thermophilic microbes)	74° 20S, 144 40E				Broady (1993) Farmer (1996) Shock (1997)

*nafter McKay et al. (1992)

The upper and middle spectra in Figure 1 have been excited using 633 and 785 nm radiation, respectively and in both cases the broad-band fluorescence emission dominates the spectrum over the wavenumber range 300-1300 cm $^{-1}$. In contrast, the spectrum excited using 1064 nm radiation in the near-infrared region shows Raman bands which can be used as biomolecular diagnostics of the protective biochemicals that are being produced by the extremophile in response to the stressed environmental conditions prevailing in the Antarctic site (Table 3). Hence, the selection of the optimum instrumental requirements and experimental conditions for the generation of Raman spectra is not a trivial task and is compounded even further by the additional constraints that are now being exercised by the proposed adoption of miniaturised Raman spectrometers for incorporation into robotic instrumentation suites for planetary landers and rovers [3]. The purpose of this chapter, therefore, is to consider the major factors which affect the observation of Raman spectra and the application of Raman spectroscopy to the characterisation of molecular entities which have relevance to the remote exploration for signs of relict or extant life in extraterrestrial scenarios.

Table 2. Sample location and habitat

	Acarospora chlorophana	*Caloplaca saxicola*	*Nostoc commune*	Cryptoendolithic lichen	*Chroococcidiopsis sp.*
Location	Football Saddle, Northern Victoria Land	Crater Cirque, Northern Victoria Land	Mars Oasis, Alexander Island	East Beacon, McMurdo Dry Valleys	Balham Valley / Barwick Valley
Latitude	72° 31'		71° 53' S	77° 50' S	77° 20' S
Longitude	169° 46'		68° 14' W	160° 53' E	161° 05' E
Altitude (m asl)	c. 800	c. 800	c.50	2200	
Distance to coast (km)	c. 10	c. 10			
Aspect, facing	S	N		N	N
Substratum, exposure and climatic factors	Pitted substratum comprising alkalibasalt, trachyandesite and phonolite. Sheltered micro-habitats on southern slope of saddle	McMurdo volcanics comprising alkalibasalt, trachyandesite and phonolite. Sheltered north-east facing slope	Aquatic and loosely settled on silt. Sheltered to south and west, within zone of enhanced UVB levels during ozone depletion	Sedimentary orthoquartzite sandstone. Exposed ridge, iron – stained outcrop	Sedimentary orthoquartzite sandstone. Sheltered valley, little iron staining
Moisture (range)	Not known. Rock surface	Not known. Rock surface	Fringes of melt pond	0.13-1.12 % dw	~1%
Habitat	Crustose	Epilithic, crustose	Aquatic and littoral	Endolithic	Endolithic

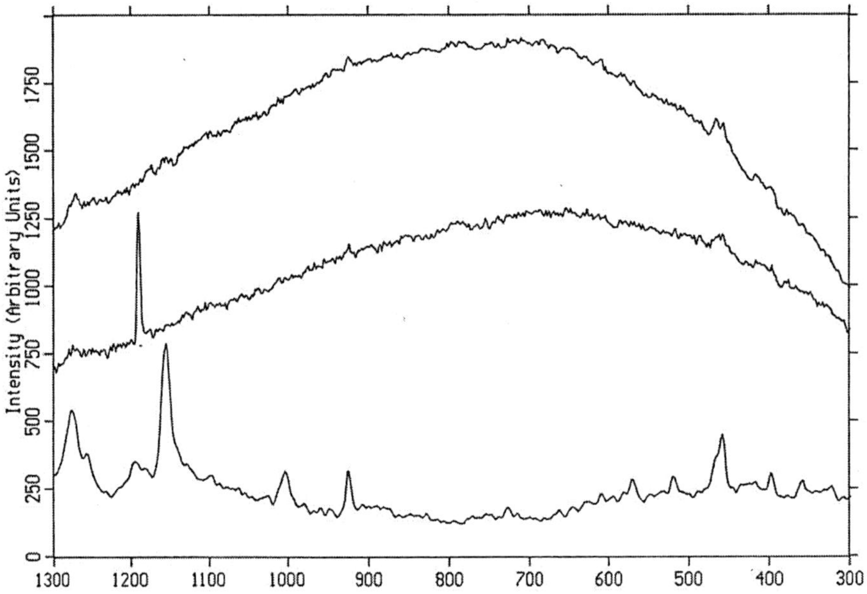

Figure 1. Raman spectra of *Caloplaca saxicola,* an Antarctic extremophilic epilithic lichen, recorded with different laser excitation wavelengths ; from top, 633 nm, 785 nm and 1064 nm. The swamping of the weaker Raman bands by the fluorescence emission background is noted for the first two wavelengths of excitation

Table 3. Raman band with assignments for protective bio-elements associated with quartz in *Acarospora chlorophana*

Excitation wavelength (nm)			Approximate assignment
1064	**852 mini**	**830**	
3069 w			ν(C=CH) aromatic
3036 w			ν(CH) aromatic
2918 s, br			ν(CH_3) asymmetric
1672 s		1672 w	ν(C=O) conjugated, free; parietin
1631 w, sh			ν(C=O) conjugated, H-bonded; parietin
1612 ms			ν(C=C) aromatic; parietin
1553 s		1554 mw	ν(C=C) aromatic; parietin
1520 vs			ν(C=C) β-carotene
1477 w, sh			Ring stretch coupled with (OH); parietin
1450 m			Ring stretch; parietin
1380 ms			Ring stretch, in plane / ν(C-O) phenyl
1370 ms			Ring stretch, in plane / ν(C-O) phenyl
1322 w			Ring stretch, in plane; parietin
1276 vs		1277 ms	Ring stretch, in plane; parietin
1254 ms, sh		1256 m	ν(C-O) aromatic ether; parietin
1198 mw			Ring stretch, C-C chelate; parietin
1180 w			Ring stretch, C-C chelate; parietin
	1161 mw, br		α-Quartz
1156 s			ν(CC) β-carotene
1137 w, sh			δ(C-CH), in plane; parietin
1105 w			δ(C-CH), in plane; parietin
	1080 vw		α-Quartz
	1062 vw		α-Quartz
1004 mw			ν(CC) ring breathing; ar. / β-carotene
926 s		927 ms	δ(C-H) out of plane; parietin
	794 mw, br		α-Quartz
	693 mw		α-Quartz
631 mw		633 m	Skeletal deformation; parietin
611 mw		611 m	Skeletal deformation; parietin
572 m		572 ms	Skeletal breathing; parietin
519 m			?; parietin
465 w, sh	470 vs*	467 m, sh	?; parietin / *α-Quartz
458 s		459 ms	δ(C=O), in plane; parietin
416 mw		423 m	?; parietin
397 m	405 mw*	400 m	Skeletal deformation; parietin / *α-Quartz
	367 w		α-Quartz
241 w		240 m, br	?; parietin

Signal strength: vs = very strong; s= strong; mw = medium; w = weak; vw = very weak; br = broad band; sharp

Factors Affecting Raman Spectral Intensity

Raman spectra are generally weak and special techniques are required for their generation and observation but the information which can be derived from Raman spectra amply repays the special attention given to their study. Raman spectra will routinely cover the wavenumber shift region from about 50 to 3500 cm $^{-1}$, which will contain all the important vibrational spectroscopic information relating to organic biomolecules and their inorganic counterparts. An example of the Raman spectrum obtained from the extremophilic endolithic cyanobacterial colonisation of Beacon sandstone (Table 1) by *Chroococcidiopsis* sp. shown in Figure 2 demonstrates this aspect very clearly. Generally, the CH stretching modes of organic aliphatic and aromatic hydrocarbon derivatives will occur [4] in the wavenumber region 2800-3150 cm$^{-1,}$ amide stretching bands of proteins near 1650 cm-^{1,} the C=C modes of unsaturated lipid chains and carotenoids between 1500-1640 cm$^{-1,}$ quadrant ring stretching bands of aromatic polycyclic hydrocarbons near 1600 cm$^{-1,}$ and various characteristic modes belonging to C-C, C-O, C-N, C-P and C-S moieties at lower wavenumbers down to 500 cm $^{-1,}$ including the uniquely observable S-S modes of cysteine residues in keratotic molecules in the region 460-520 cm $^{-1}$. At the same time the vibrational modes of molecular inorganic materials and ions such as the Fe-O in haematite, P-O in hydroxyapatite, S-O in sulfate, N-O in nitrate, Si-O in quartz and silicates such as pyroxene and olivine, C-O in carbonates, Ti-O in rutile and anatase and Fe-S in pyrites can all be characterised and identified [5] in the wavenumber range 100-1100 cm$^{-1.}$ In the particular example of *Chroococcidiopsis* sp. shown here in Figure 2, the data given in Table 4 illustrate the information that is contained in the Raman spectrum and the vibrational spectroscopic assignments that can be made of the Raman bands to the biological and geological features seen in this spectrum.

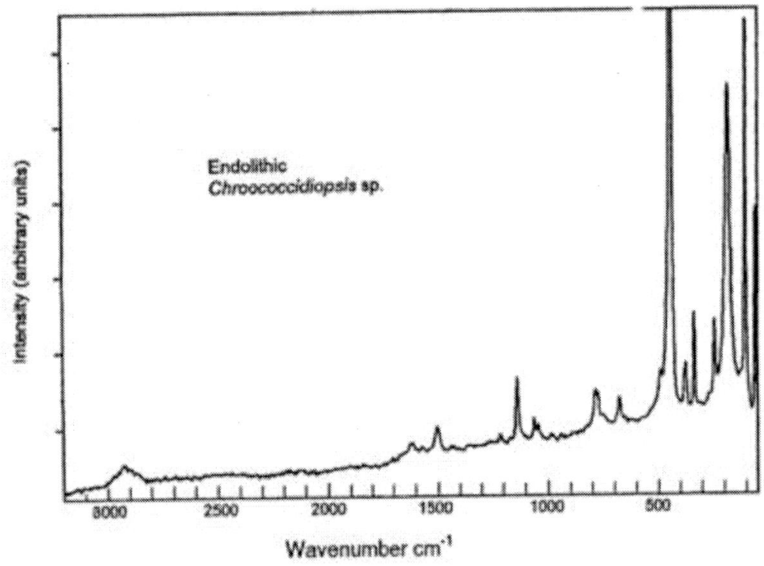

Figure 2. FT-Raman spectrum with 1064 nm excitation of *Chroococcidiopsis* cyanobacterial cryptoendolithic colony in sandstone from East Beacon, Antarctica. The full wavenumber range covers the organic Bands and those of the host geological matrix, the strongest feature being that of quartz at 463 cm $^{-1.}$

Table 4. Raman spectra band assignments for cryptoendolithic colonies in Beacon sandstone

Excitation wavelength (nm)				Approximate assignment
1064	852 port	830 substratum	514.5	
2935 m, br				ν(CH) aliphatic
1633 w, br				
1585 vw				
1521 w, br			1518 s, br	ν(C=C) β-carotene
1450 vw, br				
1230 vw				α-quartz
1159 m		1160 mw		α-quartz / ν(C-C) β-carotene
			1155 ms, br	ν(C-C) β-carotene
1081 mw, d				α-quartz
1064 mw, d		1066 w, br		α-quartz
1005 w			1006 mw	ν(C-C) aromatic
805 mw, d				α-quartz
795 mw, d		795 mw, br		substratum
695 mw		695 w		α-quartz
679 w, sh				
464 vs	467 vs	465 vs		α-quartz
400 mw, d	400 ms, br	399 mw, d		α-quartz
393 mw, d		394 mw, d		α-quartz
354 m		355 mw		α-quartz
		320 w		
263 m		264 m		α-quartz
205 ms				α-quartz

Signal strength: vs = very strong; s= strong; mw = medium; w = weak; vw = very weak; br = broad band; sh = sharp

The occurrence of both the organic and inorganic signatures in the same Raman spectrum gives a valuable and powerful analytical advantage over many other analytical techniques since now not only the presence of organics in a geological environment can be assessed but the interaction between the organics and the inorganic matrix can be examined without the necessity of extraction or separation of the phases being undertaken. Several examples of case studies will be given below to illustrate this application of Raman spectroscopy to the detection of organic molecules of relevance to life in terrestrial geological environments, which clearly have a fundamental application as a test-bed for future planned Raman analytical experiments on Mars. Other useful factors which can advocate the adoption of Raman spectroscopy as an analytical technique for potential life detection experiments in stressed terrestrial environments relate to the low Raman scattering cross-section for water and hydroxyl groups which immediately has importance for the detection of organics near deep-sea hydrothermal vents, the use of a Raman probe for molecular detection in the subsurface of glaciated lakes, the detection of life signatures in permafrost, the analysis of lacustrine sediments and for the detection of micrometeorites in Arctic and Antarctic ice-cores. In none of these examples would there be a need for specimen desiccation to record the Raman spectra, which would realistically be extremely difficult to achieve for several of the scenarios mentioned here.

There follows a consideration of the instrumental features which could dictate the selection of a particular Raman spectroscopic system, then we shall consider the use of several Raman systems in case studies to illustrate the operation of these principles in practice for the detection of biomaterials and biomolecules in real-world samples.

Wavelength of Excitation

Because of the dependence of the Raman scattering upon the fourth power of the wavenumber of the laser excitation (and inverse proportionality to the fourth power of the wavelength) the recording of Raman spectra hitherto at low wavelengths in the visible region has been strongly favoured, especially taking into account the sensitivity of radiation detectors in the visible and low wavelength regions of the electromagnetic spectrum. However, the generation of fluorescence emission, which is several orders of magnitude greater than the Raman scattering, from the population of excited vibronic states lying close in energy to the ground electronic state by high energy radiation at low wavelengths can compromise the observation of Raman spectra, as can be seen in Figure 1, for example. The onset of fluorescence emission is minimised by excitation at longer wavelengths, despite the inherently weaker Raman effect at these wavelengths, as shown in Figure 1 with the excitation effected at 1064 nm, and the manufacture of diode lasers which operate in the near infrared has created a new opportunity for the observation of quality Raman spectra which are essentially fluorescence – free from long wavelength excitation sources. This is especially useful for the recording of Raman spectra from organic biomolecules which exhibit fluorescence with short wavelength excitation, for example, porphyrins and polycyclic aromatic hydrocarbons. Another method which has been suggested for the combating of fluorescence emission in Raman spectra excited at low wavelengths is the probing of the anti-Stokes Raman spectral region, which occurs at higher energies (lower wavelength shift) than the laser line unlike the fluorescence which is a broad-band emission shifted towards longer wavelengths. However, since the anti-Stokes Raman bands are normally much weaker in intensity than their Stokes counterparts at ambient temperatures because of their dependence on the Maxwell-Boltzmann population of vibrational energy levels [2], the anti-Stokes Raman band intensities rapidly diminish with wavenumber shift from the exciting line, unless the thermal population of the excited molecular vibrational levels is increased by heating the system. Clearly, such an application, which does find practical favour in the spectroscopic determination of species temperatures in heated gases and plasmas, is not a realistic proposition for thermally cool planetary surfaces where most of the molecules will be situated in lower energy levels in the ground electronic state.

Nevertheless, the selection of a suitable laser wavelength for Raman excitation is not trivial and involves a careful consideration of the scattered Raman intensity and data acquisition over a range of selected materials; in this respect, some recent experiments have been described which inform the adoption of Raman instrumentation that will be suitable for a robotic lander on planetary surfaces.

An illustration of the effect that different wavelengths of excitation have upon the recorded Raman spectra is provided in Figures 3 and 4, which have been obtained from cyanobacterial colonisation of Antarctic rock substrates (Table 2), namely, Beacon sandstone endolith and *Chroococcidiopsis* sp., respectively. Figure 3 clearly demonstrates the inablility of the visible wavelengths of excitation to confirm the presence of geological or biological components of the specimen, whereas the Raman spectrum excited at 1064 nm provides diagnostic information about both. Figure 4 shows a similar effect for the 514.5 nm and 1064 nm excitation wavelengths, but the presence of three weak Raman features can now be seen superimposed upon the more intense background fluorescence emission at 514.5 nm ; these are three characteristic Raman bands for beta –carotene, observed at 1515, 1157 and 1007 cm $^{-1,}$ which are excited preferentially by the incident green radiation through the process of resonance Raman scattering [2], in which the intensity of certain vibrational modes involved in an electronic absorption near the excitation wavelength is enhanced by several orders of magnitude over that of normal Raman scattering.

It is also appropriate to consider the possibility of laser damage to the specimen under investigation ; a good example of this is provided in Figure 5, where the excitation of a specimen of lead (II) lead (IV) tetroxide, Pb_3O_4, is shown with *a* green (514.5 nm) and *b* red (633 nm)excitation wavelengths. The spectra are very different in appearance ; the genuine Raman spectrum is that of *b,* whereas that of *a* which has been excited with the green wavelength radiation is actually that of massicot, PbO, caused by the photochemical degradation of the Pb_3O_4 specimen by the high energy laser excitation in the green. The effects upon biological species of high energy laser radiation can be even more profound and the D and G bands of amorphous carbon at 1320 and 1590 cm $^{-1}$ are seen as evidence that extensive photolytic damage has been caused to an organic biochemical through laser irradiation at high energies. This effect is minimised using low energy radiation for the excitation of Raman spectra and must be a critical point of consideration for biological studies when evaluating laser wavelength options.

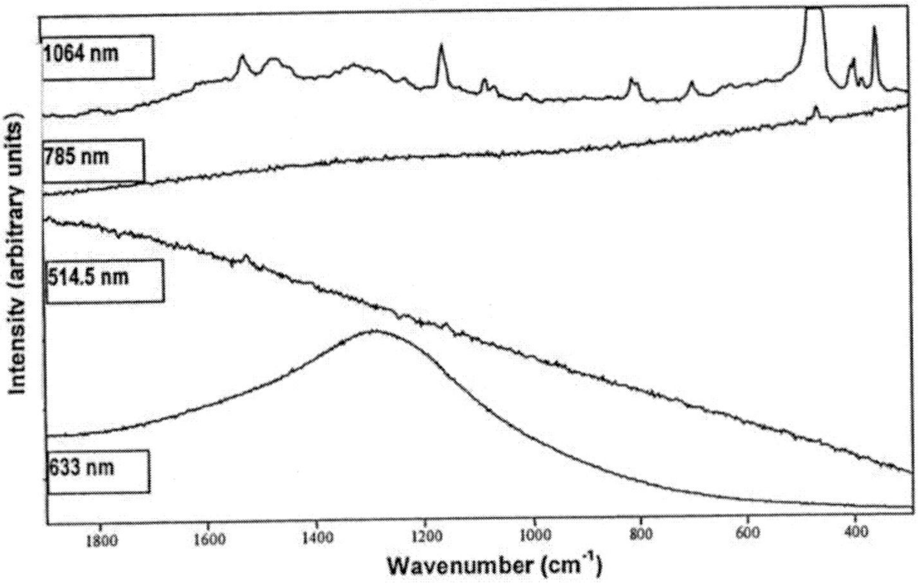

Figure 3. Raman spectral stack-plot of endolithic green algal layer in Beacon sandstone from Mars Oasis, Antarctica, recorded using laser excitation wavelengths of 514.5 nm, 633 nm, 785 nm and 1064 nm. Fluorescence emission prevents the Raman spectra of the biochemicals and of the geological matrix being shown, except for the near infrared excitation at 1064 nm.

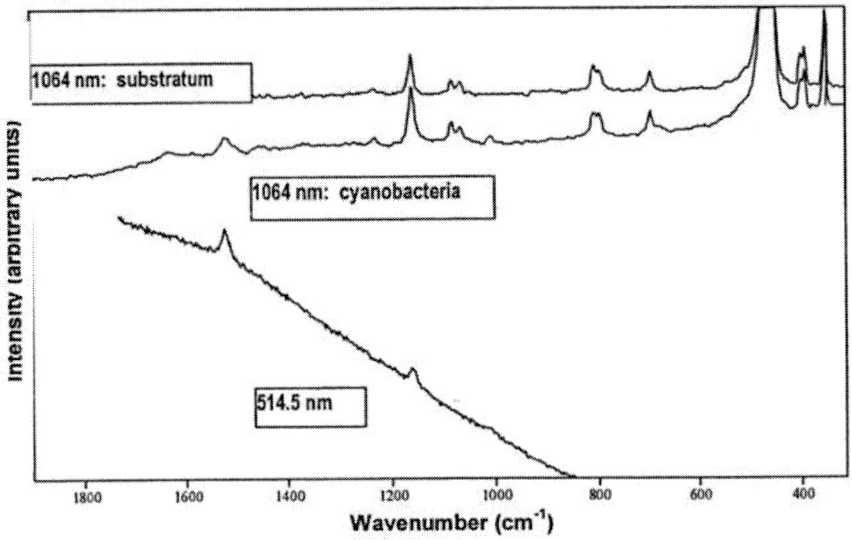

Figure 4. Raman spectral stack-plot of *Chroococcidiopsis* from Barwick Valley, Antarctic Dry Valleys, where the excitation at 514.5 nm now shows the presence of three weak Raman bands from beta-carotene at 1515, 1157 and 1007 cm^{-1} superimposed on the fluorescence emission background, compared with the essentially fluorescence –free Raman spectra obtained with 1064 nm excitation.

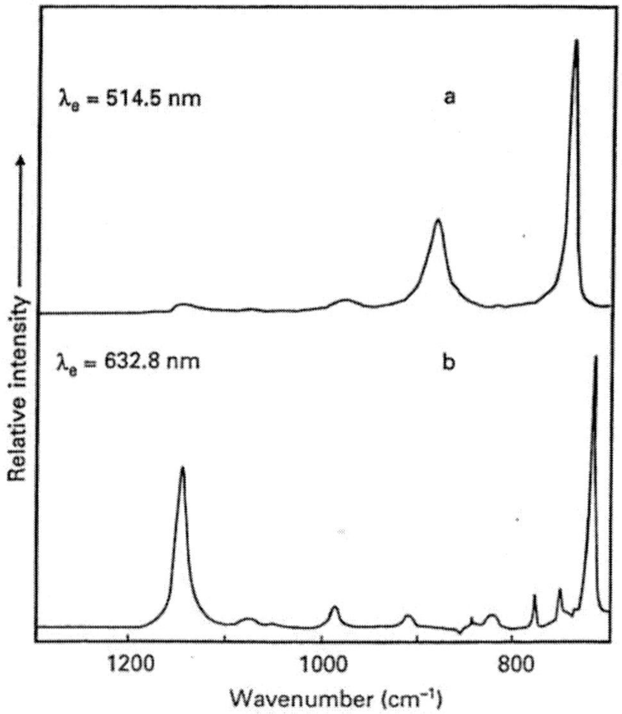

Figure 5. The effect of wavelength of laser excitation upon the observed Raman spectrum of lead(II) lead (IV) tetroxide, Pb_3O_4, excited with *a* 514.5 nm and *b*633 nm radiation. The genuine Raman spectrum is, whereas that of *a* is that of massicot, PbO, formed by degradation of the specimen.

Comparison of Dispersive-Raman and FT-Raman Spectroscopies

The point has already been made in this article that for Raman spectroscopic studies of biological materials the adoption of a system with a near-infrared excitation wavelength, for example using a Nd^{3+}/YAG laser operating at 1064 nm, obviates the problems associated with fluorescence emission that bedevils Raman spectra recorded in the blue and green regions of the electyromagnetic spectrum. A disadvantage of long wavelength excitation of Raman spectra is the fourth power dependence of the scattered Raman intensity on the wavenumber of the exciting radiation, i.e. Raman spectra excited in the ultraviolet will be significantly stronger than those excited in the near infrared. Despite this disadvantage, however, the advent of FT-Raman spectrometers operating in the near infrared at 1064 nm, which combined long wavelength laser excitation with the spectroscopic precision of interferometric measurements and Fourier transforms, provided a renaissance for Raman spectroscopic studies of sensitive biological materials which were not accessible to the technique hitherto. The development of longer wavelength diode lasers and charge-coupled detector (CCD) systems have since given additional performance attributes to conventional dispersive Raman spectroscopy, which when coupled with microscope capabilities have revolutionised Raman spectroscopy through a rapidity of data acquisition from specimen footprints of one micron or so. This has led to two-dimensional and three –dimensional Raman mapping experiments for the location of materials on heterogeneous surfaces and in transparent specimens. Whereas a detailed comparison between conventional dispersive CCD and FT-Raman techniques is often complicated by the correlation of instrumental factors with specimen requirements, the major features are as follows:

The operation of a Raman spectrometer at 1064 nm confers a reduced Raman scattering intensity over equipment that offers excitation in the visible or in the red /near infrared regions, e.g. the 633 or 758 nm wavelengths which are commonly used in CCD dispersive instruments. There is no doubt,however, that even with 633 or 785 nm excitation, some fluorescence or degradation problems in Raman spectroscopy can still occur. The major drawback to the use of CCD techniques further into the infrared is the development of stable diode lasers and effective detectors with usable Raman wavenumber shift ranges.

Despite the slower speed of data accumulation using FT-Raman instrumentation, where it is quite common to acquire data from 2000 scans, each taking one second –giving a total spectral acquisition time of about one hour, compared with a ten-spectral accumulation acquisition from a CCD-dispersive instrument, each requiring some ten seconds to achieve – giving a total spectral acquisition time of only two minutes, other experimental factors can dictate the use of the FT-Raman instrument.

In a scanning system as used by conventional CCD-spectrometers, only a specific preset wavenumber range is scanned at one time, unlike the Fellget advantage of the interferometer which means that all spectral data elements are acquired simultaneously. Hence, in the FT-Raman system, all Raman spectral data from 3500-100 cm $^{-1}$ shift are acquired in one spectral scan, which may only be taken with say 200-2000 cm $^{-1}$ for the CCD instrument. This means that careful consideration needs to be given in the latter case as to which parts of the Raman spectrum are to be accessed, since possible key vibrational indicators which lie outside the preset wavenumber range will not be recorded, whereas this does not apply to the FT-instrument that records the whole wavenumber range, and hence acquires all the vibrational information, simultaneously.

The Jacquinot throughput of the interferometric system is significantly greater than the CCD slit ; in the former ; the J stop is typically about 10 mm diameter in an interferometer compared with a slit of 100 microns in a dispersive system –in both cases, however, the effective achievable spectral resolutions are similar, being about 2 – 4 cm^{-1}.

The FT-system constantly provides an internal calibration of the data against a helium-neon laser source whereas the Raman wavenumber signals for the visible CCD-system needs external monitoring of calibration on a regular basis for spectral data correction—this is of importance for the automatic data recognition of spectral features that will be a prerequisite for remote robotic instrumentation on planetary surfaces.

A major advantage generally of Raman spectroscopy is its low sensitivity towards water and hydroxyl groups; the Raman spectrum of water is weak and much use has been made of this for the study of solutes in aqueous solutions and materials embedded in ice, for biomolecular and biopolymeric studies and for the examination of archaeological artefacts from waterlogged sites which cannot be desiccated. However, the moving of the laser excitation wavelength into the near infrared can severely restrict the Raman data that can be acquired in the presence of water because of absorption by the overtone and combination bands of water in the near infrared. Even at 1064 nm the effect of water-based absorptions upon the higher Raman wavenumber shifts can be marked and for this reason the substitution of heavy water in the form of deuterium oxide is sometimes advocated for the improvement of long-wavelength excited Raman spectra. Clearly, this is not possible for ex-laboratory and field applications but the possible resultant dimunition in quality of the Raman spectra in these circumstances should be borne in mind.

Hence, it may be concluded that for the application of Raman spectroscopic techniques to space missions the selection of a suitable wavelength of laser excitation is critically important, depending upon whether one is seeking information about the geology or the biology of specimens at a planetary surface. Also, the miniaturisation of Raman spectrometers to a size and power compatibility commensurate with a rugged and robust construction for adoption on remote landers and rovers is now advanced and several possibilities exist in a prototype stage of development. All of these involve diode lasers and simplified detectors which are tuned to operate over specified, often narrow, wavenumber ranges.

Concentration of Molecular Species

Raman band intensities are dependent upon several factors which include the irradiance of the laser ($W\ m^{-2}$) and a linear dependence on the concentration of the molecular scattering species [2]. The former factor necessitates the use of a reasonably high-powered, focussed laser beam incident upon the specimen consistent with the survival of the molecular integrity (and this can be troublesome for sensitive biomolecules which can decompose under laser irradiation) whereas the latter factor can be extremely useful for quantitative and semi-quantitative analysis of the Raman spectra. Unlike infrared optical absorption spectra, the Beer-Lambert Law relating absorbance and species concentration does not apply to Raman scattering and there can be no virtue therefore in increasing the path length of the specimen for Raman scattering to provide enhanced band intensities. The result is that Raman spectra are relatively easy to quantitate and for example, to a very good approximation, ten times the amount of material in the scattering volume will give ten times the intensity of the observed Raman band; however, this linearity of response means that there is a threshold for the observation of Raman spectra below which the observation of Raman bands is impractical. This sensitivity will be different for every molecular system and will depend upon molecular Raman scattering coefficients, which effectively describe the response of molecules to laser Raman scattering. Normally, a minor species concentration in a molecular system which represents about 0.01%-0.1% (about 100-1000 ppm) of the major component will be at the lower limit of detection in the Raman spectrum. However, as will be demonstrated below, there are several ways in which the detection of minor components in a system can be enhanced, for example using surface-enhanced Raman spectroscopy (SERS), resonance Raman spectroscopy (RRS) and a combination of these, surface – enhanced resonance Raman spectroscopy (SERRS) [6]. It has been claimed that single –molecule detection has now been achieved using the SERRS technique by coating the surface of the specimen with a specially prepared silver colloid which enhances the surface plasmon waves generating the SERS effect and judicious selection of a laser wavelength to invoke the resonance enhancement of the scattered Raman intensity [7].

Perhaps a more practical application for planetary exploration of an increased sensitivity towards the minor components in heterogeneous specimens is provided by the adoption of Raman microscopy, where the laser is focussed into a very small spectral "footprint" which may be of the order of only several microns diameter and individual particles can be examined preferentially from the surrounding matrix ; the Raman scattered intensity from these particles or small aggregates, which only represent effectively nanogram or picogram quantities of material, is enhanced by the high irradiance used in the sample illumination. Hence, good quality Raman spectra can be achieved from apparently low laser powers at source of approximately 1 mW, which imaged into a specimen particle of one cubic micron gives an irradiance of more than 1 MW m^{-2}, resulting in a Raman spectrum being obtained from only one picogram of material at a "concentration" in its matrix of less than one part-per-billion. Of course, the major problem here is the location and identification of a suitable particle to undertake the spectroscopic study! An example of the Raman spectra which can be acquired from coarse grains (ca. 0.2 mm diameter) in a heterogeneous specimen of a lacustrine sediment from Lake Hoare, an Antarctic analogue of putative Martian palaeolakes, is provided in Figure 6. Here, the diversity of geological composition is demonstrated by the very different geological signatures that are being presented in the spectra and the major mineral component in each spectrum is indicated in the figure.

Molecular Effects

The concept of molecular scattering coefficients and the sensitivity of molecules towards excitation in Raman spectroscopy has already been raised and it is appropriate here to examine the practical consequences that these parameters have upon the ability to record the Raman spectra of different materials. The sensitivity of Raman spectroscopy for the observation of Raman bands depends upon the polarisability change in electronic charge density of the chemical bonds within a molecular system on excitation with laser irradiation. Generally, bonds between large, polarisable atoms can give rise to strong Raman spectra where the number of Raman bands observed is dictated by the molecular symmetry ; the more symmetric a molecule, the fewer the Raman bands that are obtained in the spectrum. Most molecules, however, are of low symmetry or have no symmetry at all –hence, they have relatively rich Raman spectra in terms of the number of bands displayed. The wavenumber positions of the Raman bands are also dictated by the type of molecular bond that is being irradiated; Raman bands originating from bonds between heavy atoms will occur in a low wavenumber shift region of the spectrum, for example, mercury sulfide has a very strong and characteristic Raman band at 253 cm^{-1}, whereas a carbon-hydrogen bond in an aliphatic hydrocarbon will occur near 2950 cm^{-1}. This gives rise to the molecular specificity of Raman spectra, whereby it is possible to recognise the presence of molecules or chemical moieties from the wavenumber positions of their Raman bands [1].

In some cases, the Raman bands coincide in wavenumber with their infrared counterparts, but occur with significant intensity changes which arise from the different spectroscopic selection rules operating for absorption spectroscopy and scattering processes. An excellent example of this is the ketonic carbonyl, C=O, vibration which is found in proteins at 1665 cm^{-1} and is characteristic of an alpha-helical tertiary structure; although this band theoretically occurs at exactly the same wavenumber in both Raman and infrared spectroscopy, it is significantly stronger in the infrared because of the large dipole moment displayed during the bond vibration. In some cases, bands appear in the Raman spectrum, which do not appear in the infrared and *vice versa*; this occurs usually because of the operation of symmetry restrictions and is much used in the definition of molecular structures [8]. As a diagnostic device, however, the presence of key features in the Raman spectra of molecules is a very important process. Hence, the presence of a band in the Raman spectrum of keratotic proteins at about 500 cm^{-1} is a diagnostic verification of S-S bonding; furthermore, the change in wavenumber of this feature occurs in a verifiable way [9] with conformation about the C-S-S-C linkage in the protein and localised *cis, trans* and *gauche* conformational structures can be identified diagnostically between 480 and 520 cm^{-1}.

The low molecular scattering coefficient of water in Raman spectroscopy is useful for cases where desiccation of the specimen is not practicable, such as field applications in ice-laden environments or in contaminated groundwaters, such as Polar glaciated regions and volcanic geysers, respectively. The presence of water molecules in different chemical or physical sites in crystal structures is an important parameter and can be recognised through the environmental molecular effects on other neighbouring molecular ions, such as mineral hydrates, for example. Figure 7 shows the Raman spectra of two key biogeological materials that are seen to be possible signatures of extinct or extant life, whewellite (calcium oxalate monohydrate) and weddellite (calcium oxalate dihydrate), excited using 488 nm radiation. Although the water bands are not visible *per se* in these Raman spectra, the effect of the coordinated water in different sites in the crystal matrix upon the vibrational modea of the oxalate molecular ion is apparent in the wavenumbers of the stretching and bending bands of the $C_2O_4^{2-}$ ions. The recognition of these two materials in the geological record is significant fro life-detection experiments since these materials remain even after the life processes which have produced them have ceased to exist.

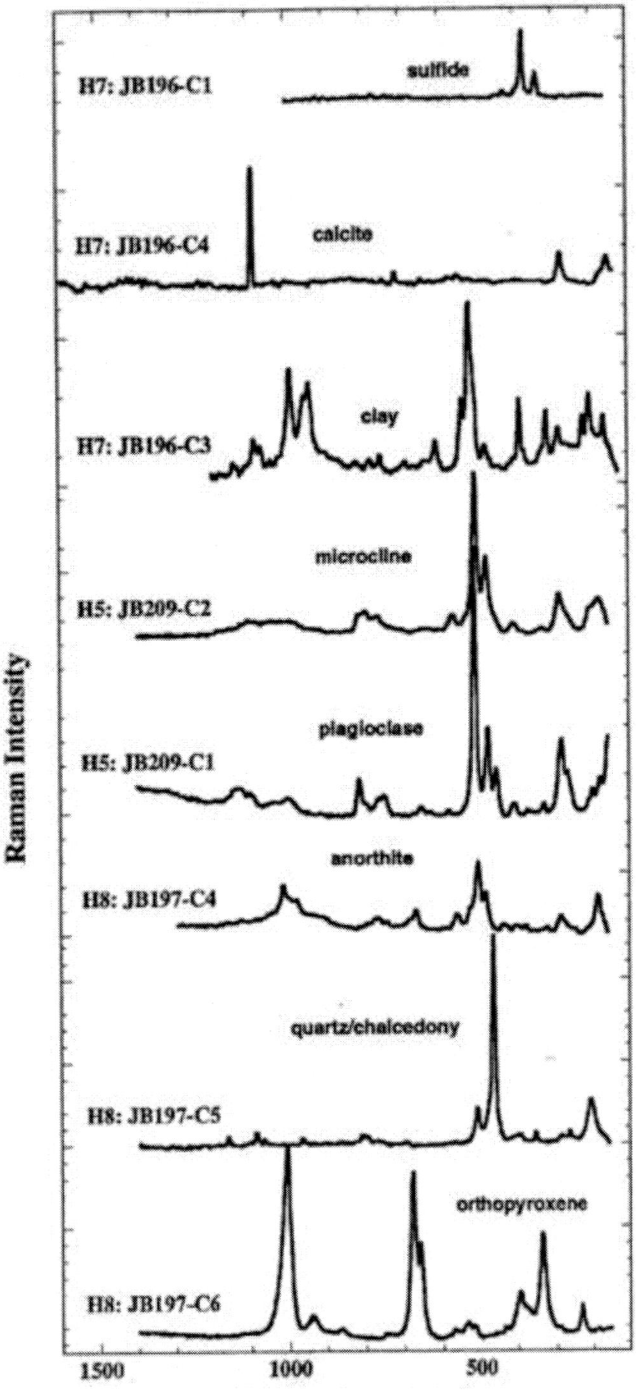

Figure 6. Raman spectra of lacustrine sediments from an anoxic region of Lake Hoare, Antarctica. The spectra are of individual grains found in three specimens, H5, H7 and H8, and the dominant mineral has been component identified in each case

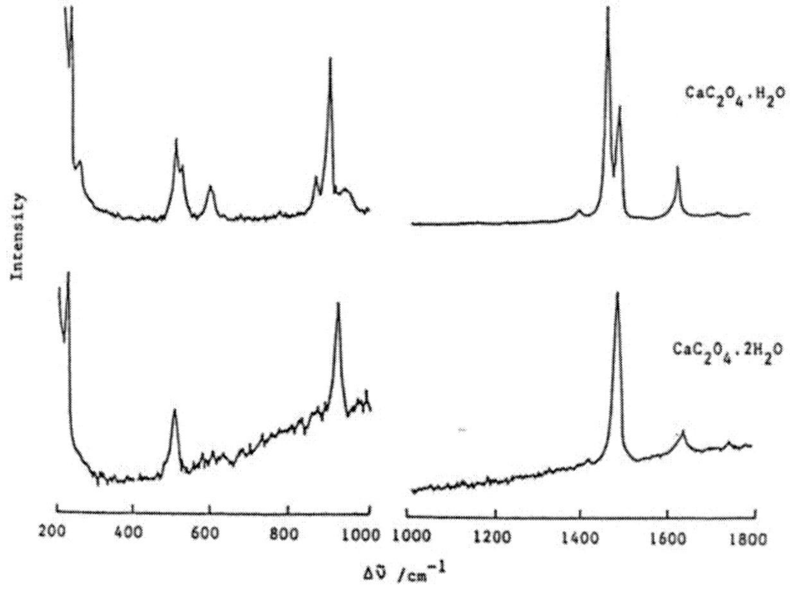

Figure 7. Laser Raman spectra of whewellite (upper) and weddellite (lower), excited using 488 nm radiation.

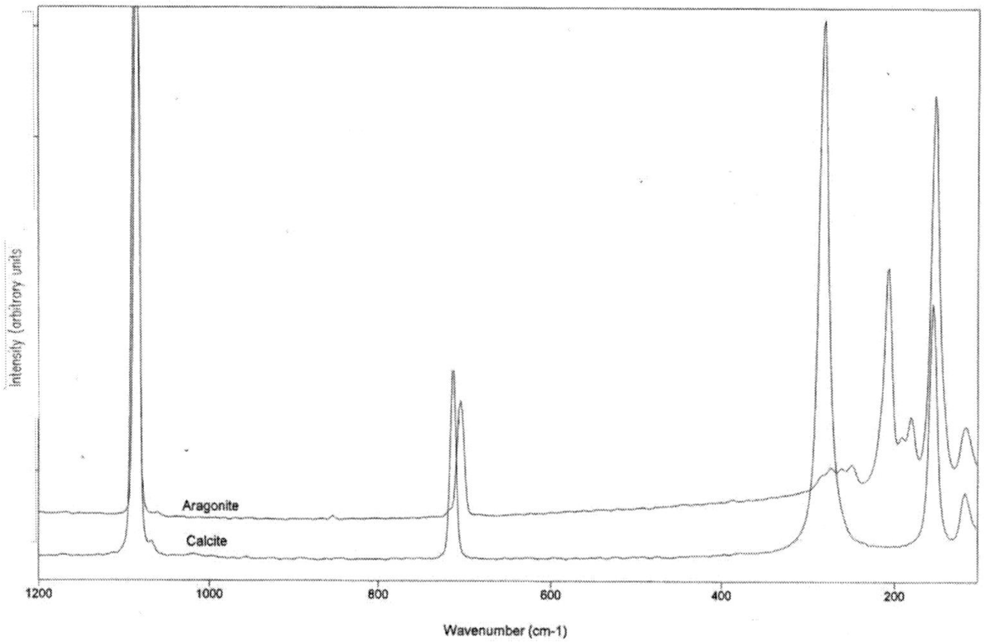

Figure 8. Laser Raman spectra of calcite and aragonite, 1064 nm excitation ; 500 scans accumulated at 4 cm^{-1} spectral resolution ; wavenumber range 100-1200 cm^{-1}.

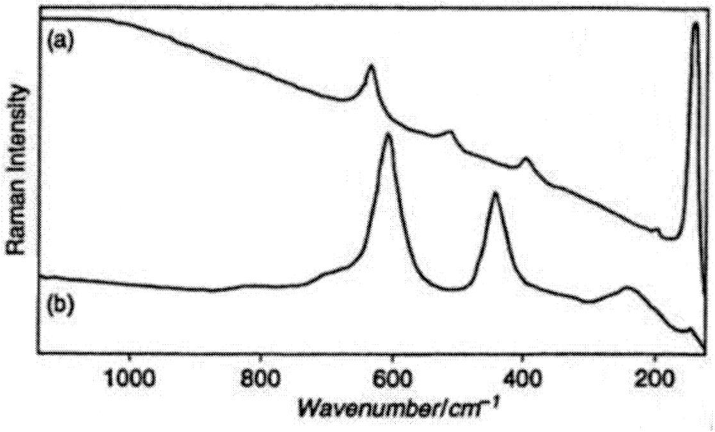

Figure 9. Laser Raman spectra of two polymorphs of titanium (IV) oxide minerals: the spectral stack plot illustrates the discrimination and specificity of the Raman data for materials which analyse as elementally identical

It should be noted that Raman spectroscopy depends upon the interactions of electromagnetic radiation with chemical bonds for the production of a spectrum. Hence, a Raman spectrum would not be expected from monatomic ions, such as Ca^{2+}, Na^{+} and Mg^{2+}. However, as has been noted in the case of the molecular electronic environments in the vicinity of water molecules, the electronic environmental effects of monatomic cations upon molecular anions can also be observed diagnostically in the Raman spectrum even when these cations themselves have no Raman spectrum. This is generally the basis of mineral diagnostic differentiation in Raman spectroscopy; in Figure 8 are shown the Raman spectra of calcite and aragonite, two polymorphs of $CaCO_3$. Here, the Raman C-O symmetric stretching band of the CO_3^{2-} ion which is common to these two minerals occurs at 1086 cm^{-1}. It is interesting that although calcite and aragonite have an identical stretching band wavenumber of 1086 cm $^{-1}$, which effectively means that this is not diagnostically useful for discriminating between these two polymorphs and can merely serve to identify calcium carbonate in a mineral matrix, the two polymorphs can be distinguished from the different wavenumbers exhibited by the other bands in their spectra, namely 712 /704 and 283 /206 cm^{-1}, respectively for calcite/aragonite. A similar situation exists for the two magnesium minerals dolomite, $CaMg(CO_3)_2$, and magnesite $MgCO_3$, where the CO_3^{2-} stretching band occurs at 1094 cm-1 for both minerals and it is now essential to use the other Raman-active bands for differentiation purposes, e.g. 713/738 cm^{-1} and 300,177,156/330,120 cm^{-1}, respectively for dolomite /magnesite. This highlights a very important point in diagnostic spectroscopy, namely that it is often not practicable to rely on just one defining band for characterisation purposes and that sometimes up to three features need to be considered. Another example of this from the area of geology and mineralogy, Figure 9, is provided by the two naturally occurring polymorphs of titanium (IV) oxide, all have characteristic Raman signatures for diagnostic bands in the spectrum, but their discrimination requires the selective observation of several features.

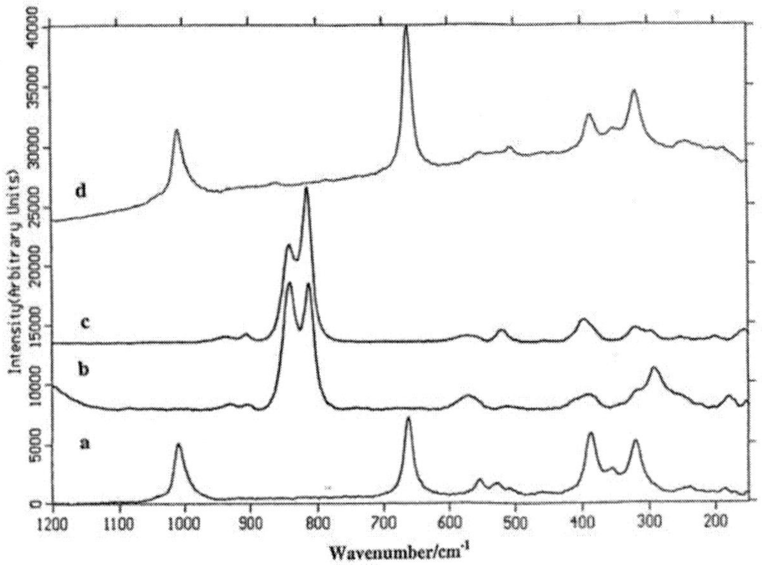

Figure 10. Raman microscope spectral stackplot of the *Nakhla* meteorite, an SNC Martian meteorite, illustrating the discrimination offered between different regions ; in each case a spectral footprint of about 2 microns is taken in the specimen and the geological heterogeneity should be noted. *a* black particle, *b* white particle, *c* brown glassy particle, *d* clear glassy particle. Raman bands at 320, 350 386, 662 and 1009 cm^{-1} are characteristic of clinopyroxene, those at 815 and 841 cm^{-1} of olivine and those at 508 and 520 cm^{-1} of plagioclase. No bands arising from biomolecules can be identified in these spectra.

The specific information provided by Raman spectroscopic analysis of geological samples directly relevant to mars can be illustrated here by the spectral stackplot of the Raman spectra recorded from the Nakhla SNC Martian meteorite shown in Figure 10. The rich diversity of chemical information that is provided by the micro-raman spectra of this specimen exemplifies the geological applications of Raman spectroscopy to Martian materials.

The construction and effective use of automatic database comparison for the recognition of Raman band signatures will therefore need to address this issue of multiple band occurrence. Although the change in vibrational band wavenumbers in the Raman spectrum can be used to provide some definitive diagnostic applications in many areas of chemical science, it does create a problem for the construction and operation of databases and especially for automated recognition of molecular entities from established databases – essentially, there is no single wavenumber position that characterises every molecule with respect to its Raman bands [8]. For example, a "free" C=C is found in the Raman spectrum at 1625 cm^{-1} for ethylene but substitution of chemical groups on the adjacent C-H bonds in the alkenyl molecule will change the wavenumber of this band, especially if the groups concerned have the ability to strongly influence the electronic environment of the isolated C=C bond through conjugation, in which the electronic density on the C=C bond is affected. A C=C-C=O moiety, therefore, will exhibit a C=C Raman band at 1605 cm^{-1} and a C=O band at 1695 cm^{-1}, both shifted down from their respective wavenumber positions in their "free" molecules. Although this is expected theoretically from molecular dynamics, it would be quite difficult to design a suitable database that would immediately recognise these bands as arising from individual C=C and C=O vibrations. Taking this discussion further to the important case of carotenoids, biomolecules based upon the chemical structures of beta-carotene and its congeners with extended chain conjugation of C=C units with each other through a C=C-C=C type coordination, the C=C vibration now appears in the Raman spectrum between 1510-1534 cm $^{-1}$ depending upon the length of the chain and end-substituents [10] ; this should be compared with the case of a free, isolated C=C in ethylene at 1625 cm $^{-1}$. Clearly, some sophistication in the automatic database for recognition of this spectral signature and its attribution as a C=C mode diagnostically would be required --- and carotenoids are identified as very significant biomolecules, having a major role in prebiotic chemical evolution discussions of the origin of life in the Solar System [11]·

Particle Size

The use of microscope objective lenses of high power and numerical aperture to record the Raman spectra of heterogeneous materials has produced some significant applications for Raman spectroscopy in the field of structural analysis and molecular species identification and have compensated for the rather poor sensitivity directed at minor components in these specimens. As illustrated above, a microscope objective lens of 100 X magnification can produce a focussed laser beam of waist diameter about 1 micron; in a confocal microscopic mode, the acquisition of spectral information theoretically in addition from a one micron segment through a sample has contributed to significant applications in depth profiling because of the real foreshortening of the dimensions of the laser focal cylinder.

However, the adoption of Raman microscopy for specimen analysis can create several problems which need to be recognised:

The high power densities at the specimen, where irradiances of MW m^{-2} can easily be realised ; whilst contributing effectively to the scattered Raman scattered intensity this can also result in specimen damage or alteration – a good example of this is provided by the change of goethite (FeOOH) to haematite (Fe_2O_3) at 633 nm with increased laser irradiance [12]. This could be very troublesome the presence of a goethite/haematite mixture on a rock surface could be indicative of endolithic colonisation in the substrata.

Particle size is a rather less well understood aspect of Raman spectra and finely divided materials. Wang *et al.* [13] observed that if the mean particle size in a specimen was greater than that of the laser excitation wavelength then the Raman band intensity decreased with the particle size. However, Hendra [14] found that the Raman intensity increased with the average particle size and with the wavelength, but if the average particle size was > 3 lambda then no significant change occurred in the Raman intensity with increase in the particle size. At particle sizes of less than 3 lambda the intensity of Raman scattering decrease with the particle size was not demonstrated. It was concluded that for a laser wavelength of 785 nm in the near infrared the limiting particle size below which the Raman intensity decreased with decreasing particle size is 2.4 microns and that as this laser wavelength is approached the Raman intensity increased for a particle size below 0.8 microns. It is clear that the influence of the particle size on the resultant Raman spectrum could be highly significant for the examination of the fine debris or detritus from drilling operations carried out using a planetary lander and that this now needs to be evaluated under terrestrial test conditions.

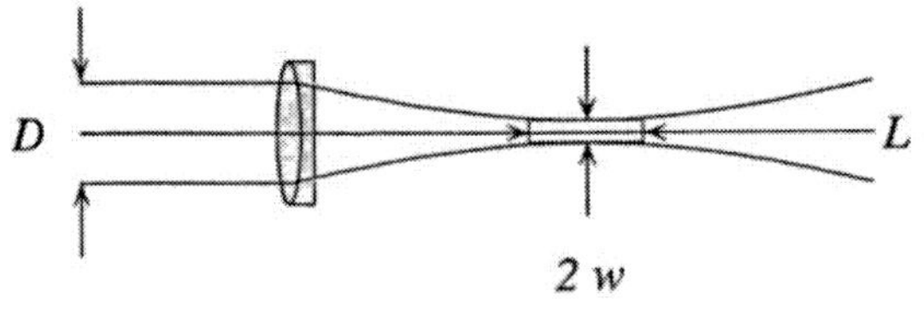

Beam waist at focus $2w = (1.27\lambda)(f/D)$

Depth of focus $L = (2.54\lambda)(f/D)$

Figure 11. Laser focal cylinder with key dimensions

Laser focal dimensions

The Raman spectra obtained from a specimen are collected not from a point source but rather from a focal cylinder which may be defined as follows [15]:

Length of waist = *w* = 0.64 lambda f/D

Depth of focus = *l* = lambda f/D

Where the numerical aperture of the objective lens and its magnification, the focal length and the wavelength of the laser radiation define the *l* and *w* parameters of the focal cylinder, Figure 11. It should be realised that some scattered radiation will be originating from out-of-focus planes above and below this focal cylinder, normally dependent upon a $1/e^2$ integration factor.

Tabaksblat *et al* [15] found that for a 40 microns particulate sample some 20 % of the total radiation collected through a 300 microns pinhole for an incident laser wavelength of 514.5 nm came from scatter planes that were > 5 microns from the focal cylinder waist plane. This measurement has given rise to much controversial argument about the actual dimensions of the focal cylinder in Raman microscope spectra which have crucial relevance particularly for the concept of *confocality.*It is now generally accepted that the definition of microscopic sample location and dimensions in confocal microscopy is not straightforward. Wijnaendts van Resandt *et al.* [16] found that for 300 mm lens objectives of 50 X and 100 X magnifications the actual lengths of the focal cylinders were 6 and 3 microns, respectively, exactly twice the values predicted theoretically. There followed much discussion [17, 18, 19] about the factors that were responsible for this inability to realise practically the theoretical optical definition of confocality. Everall [19] attributed the discrepancy between the theoretical and practically achievable confocal dimensions in Raman microspectroscopy to the effect of refraction at the surface of the specimen. He defined the true depth of focus as:

$$\text{Depth of focus} = \text{Delta}\ \{ (NA^2 (n^2-1)/(1-NA^2) + n)^{1/2} - n \}$$

Where Delta is the distance below the sample interface of the lens focus, n is the refractive index of the specimen and NA the numerical aperture of the lens. The effect of refraction at the air-specimen interface causes the focal point to be greater that previously assumed and results in a poorer axial resolution (depth) than predicted theoretically. Another factor that emerged from this treatment was that the depth resolution also was predicted to change with the focal distance and depth within the specimen itself. From measurements on a silicon wafer, Everall[19] showed that a calculated 2 micron depth of focus actually gave a true depth of focus (focal plane) of 2.2 Delta, which evaluates as 11 microns for 514.5 nm laser radiation.

In a subsequent article Everall [20] extended and modified his original calculations which then were confirmed with the findings of Baldwin and Batchelder [21] that the depth of focus calculations were under estimated for larger specimen sections but were realistic approaches for thin films. The real situation is even more complex when the distortion produced by spherical aberration mismatch in the refractive index of the specimen resulting in a focal smearing occurs; this is very dependent upon the numerical lens aperture and is pronounced for large numerical apertures [22, 23]. The end result is that one should not be too definitive about the precision of the depth of focus in a confocal Raman microscopy experiment and that the Raman spectra recorded from subsurface layers in transparent materials are subject to "spillover" from adjacent layers which are not formally considered to be significant theoretically butcertainly can be critical experimentally.

Spectral Resolution

The effect of the instrumental spectral resolution upon the appearance of the resultant Raman spectra with regard to the capability of discrimination between the biological and geological band signatures is a critical one. Although the spectral resolution, measured in cm $^{-1}$, is operator adjustable in laboratory instrumentation it will normally be a fixed parameter consequent upon the design features in the case of miniaturised instruments. The effect of changing the spectral resolution for 1064 nm laser excitation upon the Raman spectrum of olivine illustrates this point. The characteristic doublet at 820 and 860 cm $^{-1}$ using a spectral resolution of 4 cm^{-1} provides a key diagnostic indicator for this mineral. At a spectral resolution of 32 cm^{-1}. However, this doublet collapses to a broader singlet band with a peak wavenumber of 844 cm$^{-1.}$ The result of the recording of the Raman spectrum of olivine at this lower resolution of 32 cm^{-1} is that a database search mechanism would probably fail to assign the specimen to olivine. Figure 14 shows a similar situation applies for biomolecular analysis, namely an epilithic lichen, *Acarospora* sp. form Vegetation Island, Antarctica, excited using 1064 nm radiation with a spectral resolution of 4 cm $^{-1}$ and also acquired with a spectral resolution of 32 cm $^{-1.}$An even more serious situation can now be envisaged than was the case for the geological specimen of olivine considered earlier, in that many more key biological signatures that are essential for the identification of the complex biomolecular composition of the organism are now obscured at the lower resolution; the key features due to rhizocarpic acid at 1664, 1595 and 1001 cm $^{-1}$ are seen at higher resolution, but with poorer resolution data is observed at 1663 and 1599 cm $^{-1}$ only. The sharp band at 1001 cm$^{-1,}$ a key feature for monosubstituted aromatic ring compounds, is now totally absent at lower resolution. Similarly, the quadruplet in the neighbourhood of 1300 cm $^{-1}$ has now collapsed to two broader bands at 1346 and 1287 cm $^{-1}$, from 1347 and 1280 cm^{-1} in the spectrum recorded at higher spectral resolution. The bands at 1159 and 1520 cm $^{-1}$ from a carotenoid component are also absent at lower resolution, as are features at 944 and 787 cm $^{-1}$.

Clearly, the operating resolution under which the Raman spectra are being recorded will be vital for the effective database discrimination capability that will be an essential component of miniaturised spectrometer design. Instrumental evaluation of proposed miniaturised spectrometers must be an integral part of their incorporation into planetary landers. As the instrument size is diminished physically, each pixel in the diode array detector covers a larger wavenumber range and a choice facing instrument constructors is the compromise between wavenumber shift range versus the spectral resolution -- the former will decrease the spectral coverage for materials of interest whereas the latter can distort the observed spectra which may result in a mis-attribution of the data or even missing features, the presence of which may be vital particularly for the biological assignment.

Adaptation to Mars

Rovers have revitalised the possibility of finding the signs of life on Mars –extinct or extant. We now have evidence of images of a planetary surface [24, 25] that clearly has suffered water erosion, subsurface aquifers, the possibility of subglacial polar lakes (similar to Ellsworth and Vostok in Antarctica) and a tantalising glimpse of geological niche scenarios that match terrestrial models, for example Mars Oasis, Antarctica [26] (Figure 12), where biological adaptation and survival to these harsh Martian environments may have occurred. The new phase for Martian exploration now firmly rests on the molecules that have been synthesised by extremophiles for their protection against low wavelength radiation and desiccation and the associated auxiliary molecules that are required to harness the limited energetics and nutrients that are available in their protected niche environments [27, 28, 29]; finally, in a "limits of life" scenario, where life has become extinct, the remote analytical observation of residual molecular signatures in the geological record may perhaps indicate that the evolutionary experiment on Mars commenced but was not successful in its survival (Figure 13).

Extremophiles

Extremophiles are organisms that can survive where humans cannot; their survival in hostile terrestrial environments depends critically upon their ability to synthesise sophisticated suites of protectant biomolecules (Table 5) and the adaptation of geological and hydrological niches for colonisation to combat low wavelengths of insolar radiation, desiccation,and extremes of temperature, pH, salinity, pressure and heavy metal toxins [29,30]. The search for life on Mars is linked closely to the discovery of novel strategies adopted by extremophiles in terrestrial scenarios where previously the existence of life was believed impossible. A selection of extremophiles and their terrestrial habitats will be considered here, which has provided astrobiologists with the term *"terrestrial Mars*

Figure 12. Mars Oasis, Antarctica; a terrestrial niche for the survival of endolithic cyanobacteria in a "*limits of life*" situation; a translucent Beacon sandstone outcrop containing endolithic microbial communities on top of a dolerite sill. Dr David Wynn-Williams, Head of Antarctic Astrobiology at the British Antarctic Survey, Cambridge, is prospecting for endoliths. Reproduced from the book Astrobiology: The Quest for the conditions of Life, Eds: Gerda Horneck, Christa Baumstark-Khan, 2003, Springer. With kind permission of Springer Science and Business Media.

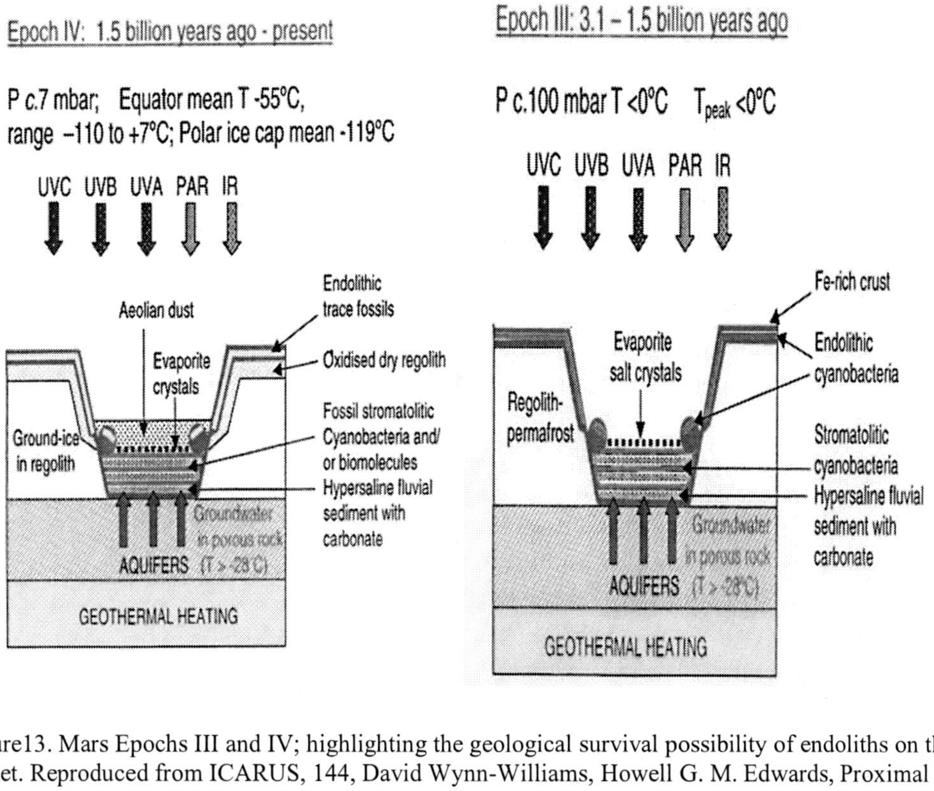

Figure13. Mars Epochs III and IV; highlighting the geological survival possibility of endoliths on the planet. Reproduced from ICARUS, 144, David Wynn-Williams, Howell G. M. Edwards, Proximal Analysis of Regolith Habitats, p486-503, 2000, with permission from Elsevier

analogues" on which the experimental remote analytical projects envisaged for robotic Mars explorations are now being evaluated. The driving force behind the detection of life signatures on Mars is really two-fold, in that the search for a biological evolutionary occurrence other than our own will undoubtedly enhance the definition of our own early prebiotic chemistry but it will also engage with the efforts of space exploration in the placement of a human being on the surface of Mars, which will require a fuller knowledge of the planetary composition and its potential biological dangers than we have currently [31, 32].

In this respect, the study of extremophile survival strategies which operate in some of the most inhospitable places on Earth will further our understanding of our own precarious existence in a changing natural world.

Extremophiles that have been studied in our laboratories and which address several scenarios that could be expected on Mars, i.e. *terrestrial Mars analogues*, include the following:

- Life in rocks: Antarctic endolithic microbes;
- Shocked rocks: pore colonisation;
- Subglacial lakes: polar colonisation.
- Meteorite impact craters: halotrophic cyanobacteria;
- Lacustrine sediments: microbial colonies;
- Volcanic lava deposits: vacuole colonies;
- Stromatolites: cyanobacterial residues;

Chasmolithic Community

A chasmolithic community living in a fracture inside a marble rock was collected during the AMASE expedition (Arctic Mars Analogue Svalbard Expedition) in August 2004 to the Vest Spitzbergen Island in the Svalbard Archipelago, sited 80°N, inside the Arctic Circle. Although the temperatures are not as cold as those in Antarctica (at Spitzbergen Island, the winter temperature reaches minus 30-35 °C), the site and the samples are considered extremophiles because of the low temperatures and the absence of sunlight for almost six months each year.

The sample was found in a glacier moraine in Bockfjorden, at an altitude of 30 metres above sea level, and shows an epilithic community on the crust which gives a greyish tonality to the white stone substrate, and a chasmolith inside a crack where several coloured strata, namely pinkish, black, green and brown areas, are visible.

The Raman spectrum of the substratum presents bands at 1097 cm^{-1}, which is seen as a shoulder on the stronger signature at 1086 cm^{-1}, 713 cm^{-1} with a shoulder at 725 cm^{-1} and 157 cm^{-1}. This provides an excellent example of the use of Raman spectroscopy to distinguish between isomorphic minerals. The bands at 1097, 725, 299 and 177 cm^{-1} are characteristic of dolomite, a calcium magnesium carbonate ($CaMg(CO_3)_2$), whereas calcite, the calcium carbonate isomorph, shows signatures at 1086, 713, 281 and 157 cm^{-1}, very close in wavenumber, but readily distinguishable from the spectral signatures of the dolomite. Figure 14 gives a Raman spectral stackplot of magnesite, $MgCO_3$, with Raman bands at 1094, 738, 326 242 and 119 cm^{-1}, dolomite, $CaMg(CO_3)_2$, with bands at 1098, 725, 299 and 177 cm^{-1}, aragonite, $CaCO_3$, with bands at 1086, 704, 208 and 154 cm^{-1}, and calcite, $CaCO_3$, with bands at 1086, 713, 283 and 156 cm^{-1}.

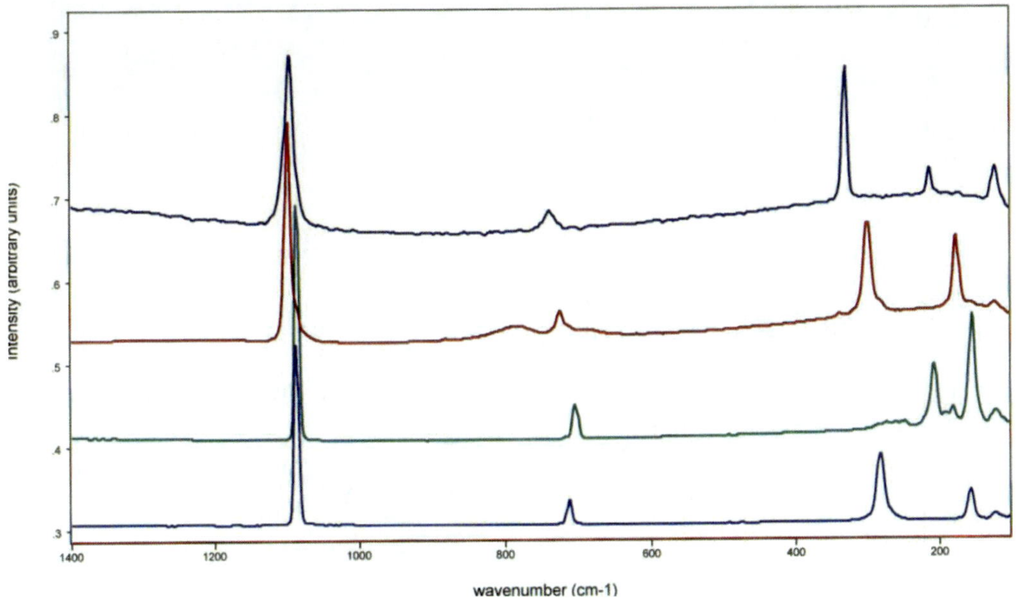

Figure 14. Raman spectra of, from the top: magnesite, dolomite, aragonite and calcite-indicating the spectral discrimination between these carbonates.

On the surface crust, the grey tonality is the result of cyanobacterial colonisation, which gives the characteristic Raman spectral bands of scytonemin. This UV-protective pigment is not found in the chasmolithic organisms inside the crack and has probably been produced as a strategic response to the increased radiation levels experienced at the rock surface.

Figure 15 shows a section of chasmolithic feldspar from Lake Hoare, Antarctica, in which two zones of bacterial colonisation can be clearly seen, from which the Raman spectrum of calcite can be clearly identified closely associated with the colonisation zones signatures of quartz and feldspar can also be seen in the spectrum.

On the surface crust, the grey tonality is the result of cyanobacterial colonisation, which gives the characteristic Raman spectral bands of scytonemin. This UV-protective pigment is not found in the chasmolithic organisms inside the crack and has probably been produced as a strategic response to the increased radiation levels experienced at the rock surface.

Figure 15 shows a section of chasmolithic feldspar from Lake Hoare, Antarctica, in which two zones of bacterial colonisation can be clearly seen, from which the Raman spectrum of calcite can be clearly identified closely associated with the colonisation zones signatures of quartz and feldspar can also be seen in the spectrum.

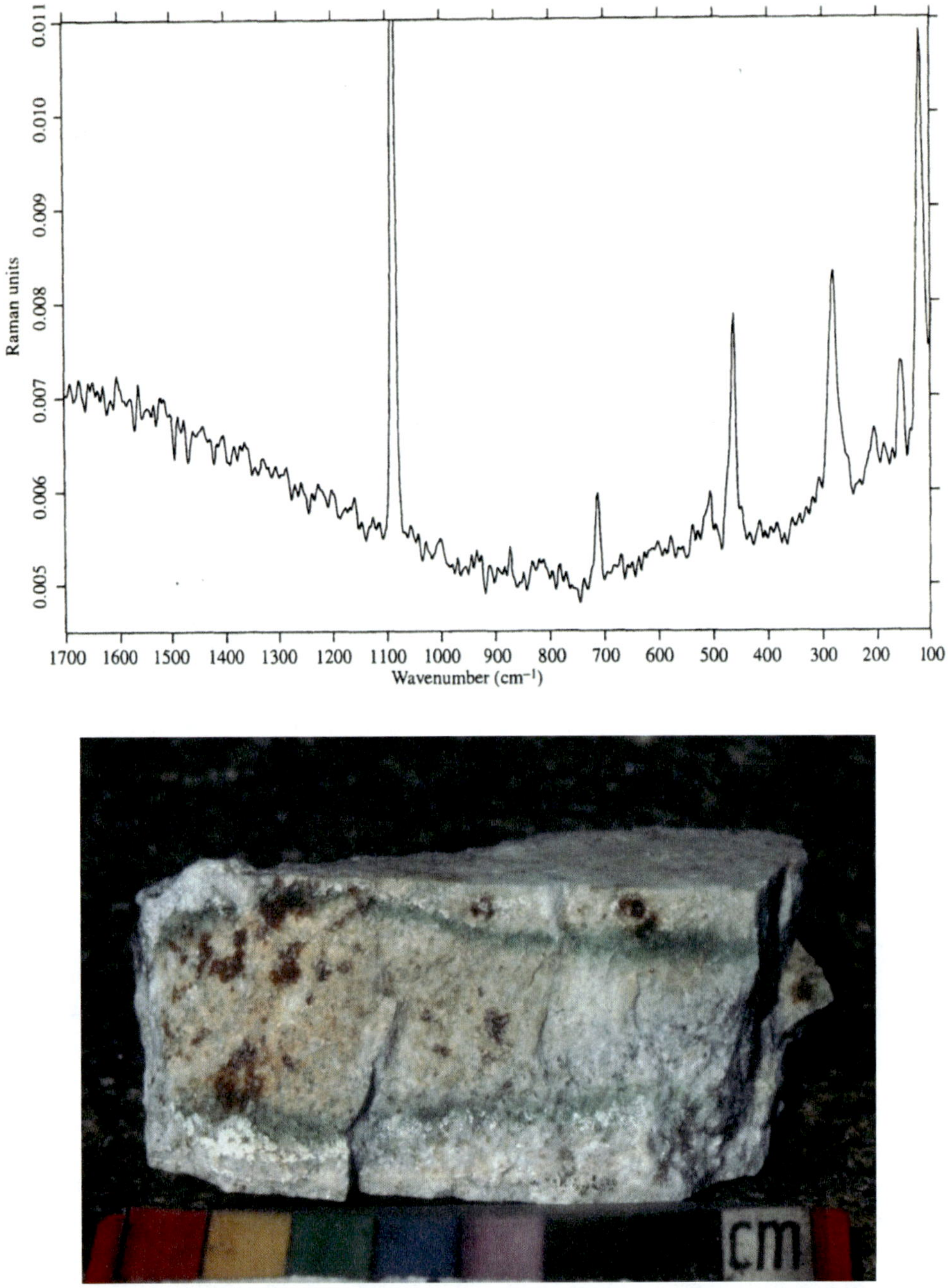

Figure 15. Chasmolith in feldspar, Lake Hoare, Antarctica: with the Raman spectrum of calcite, quartz and feldspar from the cyanobacterial colonisation zone

Table 5. Raman spectral signature bands and UV-visible absorption maxima for functional microbial pigments in extreme terrestrial Antarctic habitats

Physiological Function	Pigment	Pigment Type	Raman vibrational bands (wavenumber cm^{-1})					Absorption peaks (nm)			Visible >400	
								UVC <280	UVB 280-320	UVA 320-400		
Ultraviolet Screening pigments	Usnic acid	Cortical acid	2930		1607	1322	1289	>220	290	325	<400	
	Pulvinic dilactone	Pulvinic acid derivative		1672		1405		>246	290	367		
	Paretin	Anthraquinone		1675		1099	551	>257	288		431	422
	Calycin	Pulvinic acid derivative			1611		1379	269				
	Atranorin	*para*-depside	2942	1666	1303	1294	1266	<274			>400	
	Gyrophoric acid	Tri-depside		1661		1290		275	304			
	Fumarprotocetraric acid	Depsidone		1642	1630	1290	1280	273	315			
	Emodin	Quinone		1659					291		440	
	MAA *(Nostoc)* 7437	Mycosporine Amino Acid	2920		1400		820		(>310)	330		
	Scytonemin	8-ring dimmer		1590	1549	1323	1172	252	300	370	>400	
Anti-oxidants Photo-synthesis	β-carotene	Carotenoid			1524	1155		<246	283	384	429	
	Rhizocarpic acid	Isoprenoid		1665	1620	1596		>200				
	Porphyrin	Tetrapyrrole ring			1453							

Table 5. (Continued)

Physiological Function	Pigment	Pigment Type	Raman vibrational bands (wavenumber cm^{-1})					Absorption peaks (nm)			Visible >400	
								UVC <280	UVB 280-320	UVA 320-400		
	Chlorophyll, (Cyano bacteria)	Tetrapyrrole ring			1360	1320					680	700
	$BChl_a$ (*Rhodopseudomonas*)	Tetrapyrrole ring	na								850	870
	Chlorobium Ch1	Tetrapyrrole ring	na								650	660
Accessory Light-harvesting												
	Phycocyanin	Phycobilin		1638	1369						560	620
	Phycoerythrin	Phycobilin	na								544	

In the chasmolithic pink coloured zone area, the Raman spectra show bands of chlorophyll and a carotenoid, but the fluorescence emission is very strong here, which reduces the quality of the Raman signatures. The green layer gives a strong fluorescence emission with 785 nm laser excitation and no spectral signatures are visible, but selection of 514 nm laser excitation provides bands from a carotene which are clearly recognizable at 1524, 1157 and 1000 cm^{-1} (Figure 16); these signatures are consistent with an assignment to astaxanthin. The same applies to the black zone in the chasmolith, but now the carotene gives bands at 1516, 1154 and 1004 cm^{-1} and so it can be identified as beta-carotene. Although no spectral signatures could be recorded for this system with the near-infrared laser at 1064 nm, three different micro-organisms are differentiated with 514 nm excitation because of the different carotenes produced, two discrete organisms occurring in the green band the third in the black band.

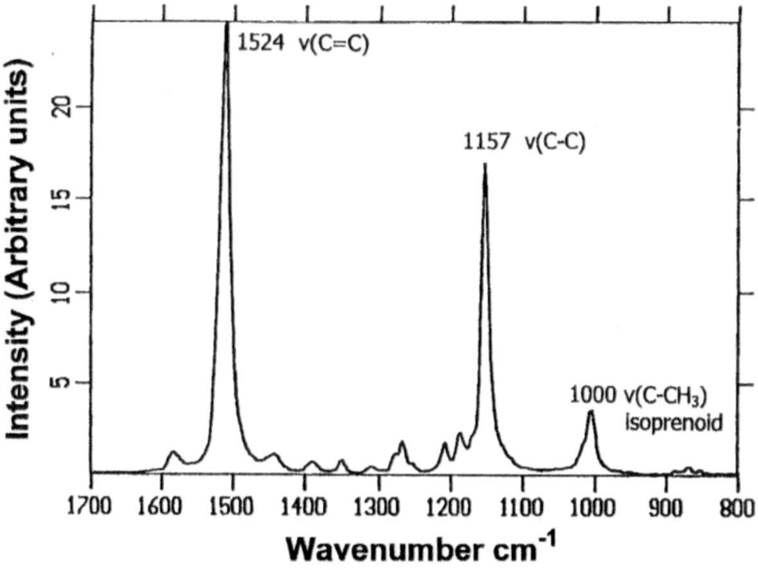

Figure 16. FT-Raman spectrum of a carotene found in an Antarctic Beacon sandstone endolith, Battleship Promontory, Alatna Valley.

Endolithic Community

This specimen was collected at Mars Oasis in Antarctica, the coldest desert on Earth, during the summer season of 2002. The sandstone shows a normal orange-red colouration but a depletion of the red mineral (haematite) is clearly visible in the vicinity of the endolithic organism (Figure 17).

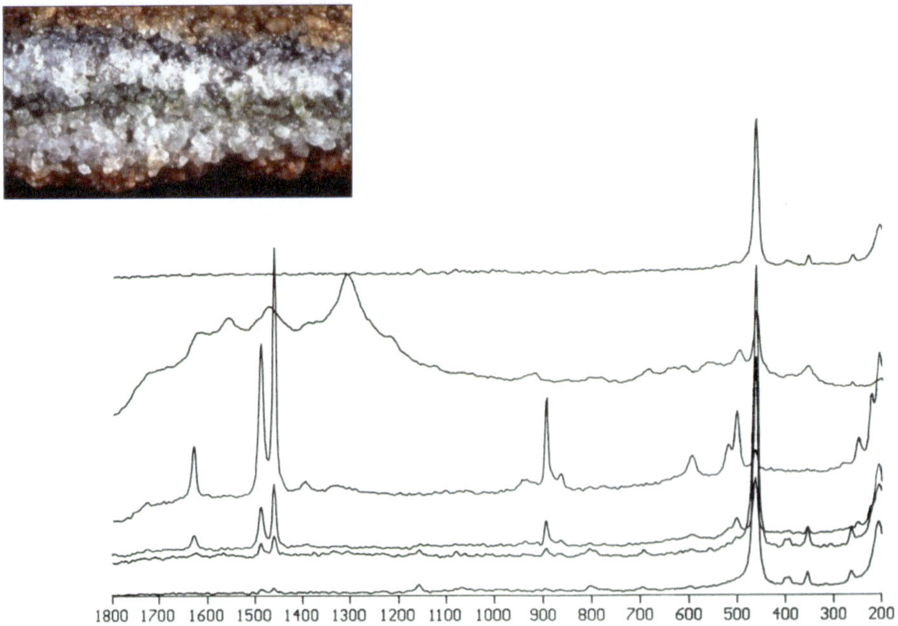

Figure 17. Endolithic colonisation of Beacon Sandstone, Mars Oasis, Antarctica, showing zonation. The lower interface of the algal zone dominated by cyanobacteria occurs 8 mm below the upper surface of the rock.

Quartz is the main component of the rock, with signatures at 128, 206, 355, 542, 696, 795, 807, 1064, 1081, 1161 and 1227 cm^{-1}; the red colour is recognizable as haematite because of its Raman bands at 223, 291, 404, 495 and 609 cm^{-1}. This is abundant below the organism stratum but has been removed from the immediate vicinity, probably to permit photosynthetic-active radiation to reach endnolithic community. The specificity of Raman spectroscopy permits one to distinguish the very weak signature of haematite in the crustal spectrum (Figure 18); haematite has been proved to be a UV-screening compound and it is interesting that the organisms can maintain this mineral on the crust for the purpose of protecting against radiation that reaches the terrestrial surface (exacerbated by the depletion of Antarctic ozone). This particular crust colouration, modified by the underlying biological colony, was used by the prospecting geologist as a visual clue to the presence of endolithic communities inside the rock.The stackplot of spectra collected from the endolithic specimen shows bands of calcium oxalate monohydrate at 1490, 1463, 896, 502 and 206 cm^{-1}in the depleted pigment area. Two different calcium oxalates are identified by Raman spectroscopy because of their signatures in extremophile exemplars; whereas in the calcium oxalate monohydrate (whewellite) Raman spectrum, apart from some weaker signatures, two bands are shown at 1490 and 1463 cm^{-1}, for calcium oxalate dihydrate (weddellite) only one characteristic strong signature, centred at 1475 cm^{-1}, is visible. Even when they appear in admixture whewellite and weddellite are clearly distinguishable using the Raman spectroscopic analytical technique. In addition, beta-carotene with bands at 1516, 1154 and 1004 cm^{-1} and chlorophyll, with a strong and broad signature at 1327 cm^{-1} and weaker signals at 915, 744 and 517 cm^{-1}, have also been identified, as well as quartz.

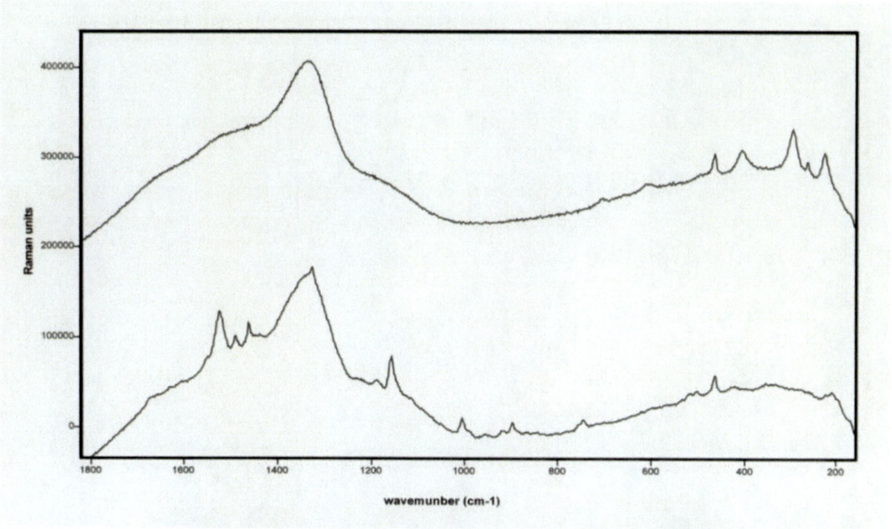

Figure 18. Raman spectrum of haematite crust on sandstone endolith. The Raman spectral stackplot shows the spectra obtained from a vertical transect through the surface crust, the biological hyaline layer, algal layer, depletion zone and basal rock, from the top.

Volcanic Rock Colonisation

A primary colonization in fumarole minerals from a recently active and still warm (two week-old) lava flow was collected by the author during a field trip to the Kilauea volcano, Hawaii, in 2004 and was analyzed using Raman spectroscopy about seven days after collection, figure 19. Besides identifying the mineral components, a carotenoid was recognized because of the Raman signatures at 1530 and 1143 cm^{-1}, and a chlorophyll-like compound could be related to the signatures at 1449, 1342, 952, 834, 748, 680, 595, 438, 257 and 173 cm^{-1}. This result indicates that the organisms needed only a very short time indeed to colonize an inhospitable niche such as a fumarole in a still very warm and active lava flow.

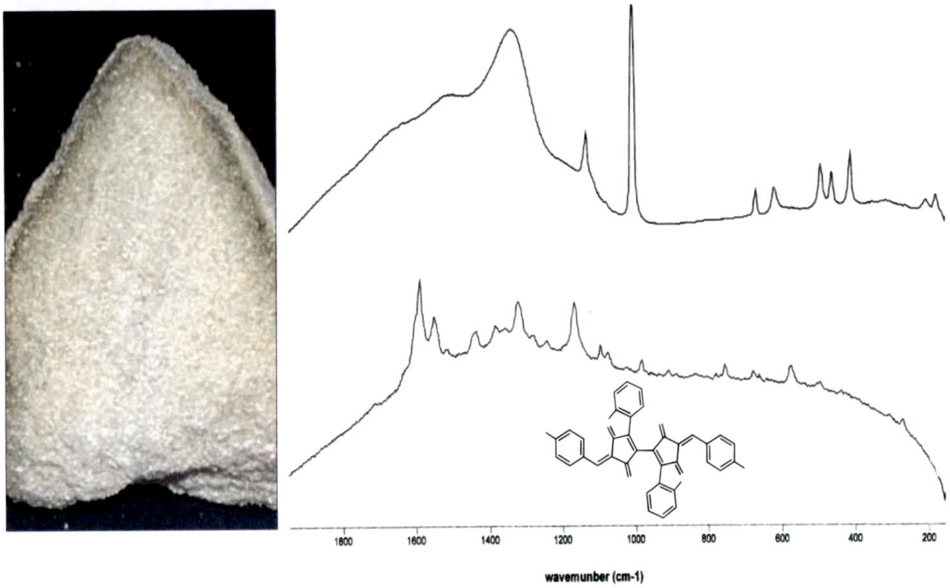

Figure 19. Antarctic endolith in gypsum, with Raman spectra of, upper spectrum, the surface (gypsum) and lower spectrum, the underlying cyanobacterial zone (scytonemin). The chemical structure of scytonemin is also given here.

A second sample of lava (Figure 20) was collected during the AMASE expedition of 2004 in Scott Keltiefiellet, in the Vest Spitzbergen island of Svalbard [40] at 80°N at about 1,000 m altitude. Organisms were found inside a vacuole of about 2 cm diameter in a black lava basalt matrix; the vacuole possessed a simple orifice of less than 1 mm diameter through which the organisms, sun, water and gases could penetrate. Technically, endoliths are organisms living inside the pores of a sedimentary rock, but this new kind of "endolithic" community affords a novel opportunity to look for life inside volcanic rocks that would not normally have been considered to be a geological niche for endolithic colonisation.

Three different carotenes were observed in the vacuole when 514 nm laser excitation was used, but there were no chlorophyll signals. With 785 nm excitation only two carotenes were detected but chlorophyll signals appeared alongside the carotenes and characteristic bands from c-phycocyanin, an accessory light-harvesting pigment, were also detected.

Halotrophs

The presence of salts in SNC (Shergotty-Nakhla-Chassigny) Martian meterorites [32-35] and the acknowledgement of the presence of sulphates in the Martian regolith [35-38], together with the recognition of the existence of lacustrine water sediments at or below the surface of Mars [39-44] makes terrestrial halotrophic organisms potential Martian analogues.

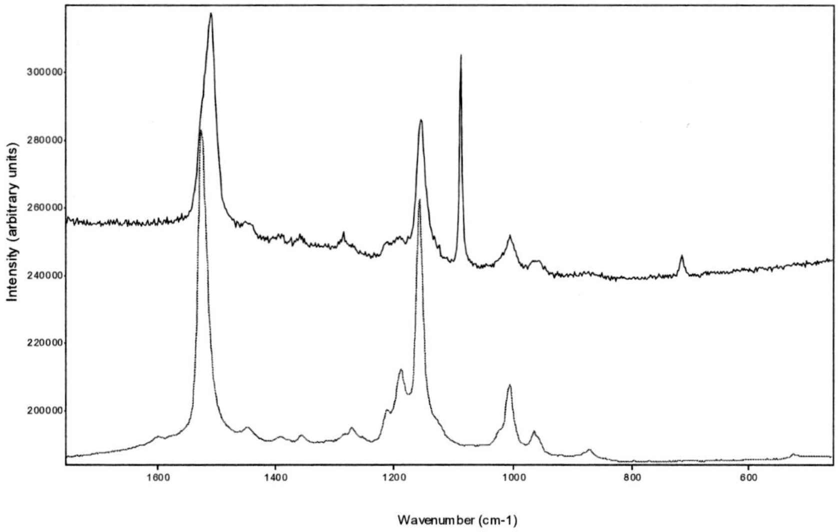

Figure 20. Novel endolith in a vacuole in a volcanic basalt lava matrix, Svalbard, Spitsbergen, Norwegian Arctic, showing two different carotenoids in admixture. Accessory photopigments are also identified in this system (not shown here).

Apart from the Raman spectroscopic identification of two different types of bacteria in a gypsum crystal (*Nostoc* and *Gloeocapsa*), it was also possible to detect organic signatures from bacterial colonies sited several millimetres below the surface in a transparent crystal of selenite from a 26 Mya meteoritic impact crater [38] at Devon Island in the Canadian High Arctic (Figure 21).

The Raman spectrum of the selenite crystal surface shows the strongest gypsum band at 1006 cm^{-1} and other minor gypsum signatures; Raman spectra taken from a vertical transect through the crystal from the surface, using a confocal laser beam to obtain spatial information from within the crystal, provides evidence of additional features in which the Raman bands, can be identified with the cyanobacterial signatures. Confocal laser probing of the dark-coloured inclusion within the crystal reveals that it is a halotrophic cyanobacterial colony containing the radiation protectant scytonemin (Figure 22).

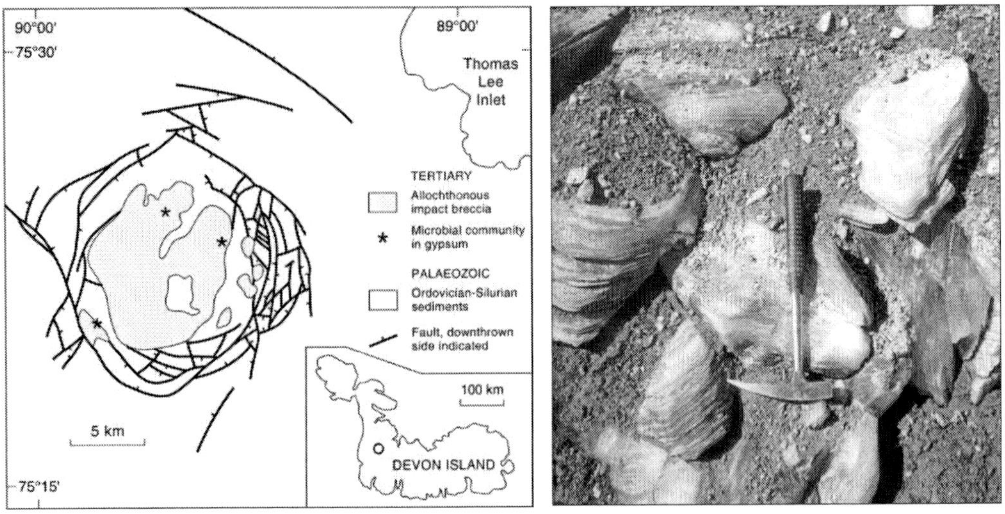

Figure 21. Selenite var. gypsum) deposits, Haughton meteorite impact crater, Devon Island, Canadian High Arctic, 26 Mya. Also shown are details of selenite crystals within surface melt breccia deposits.

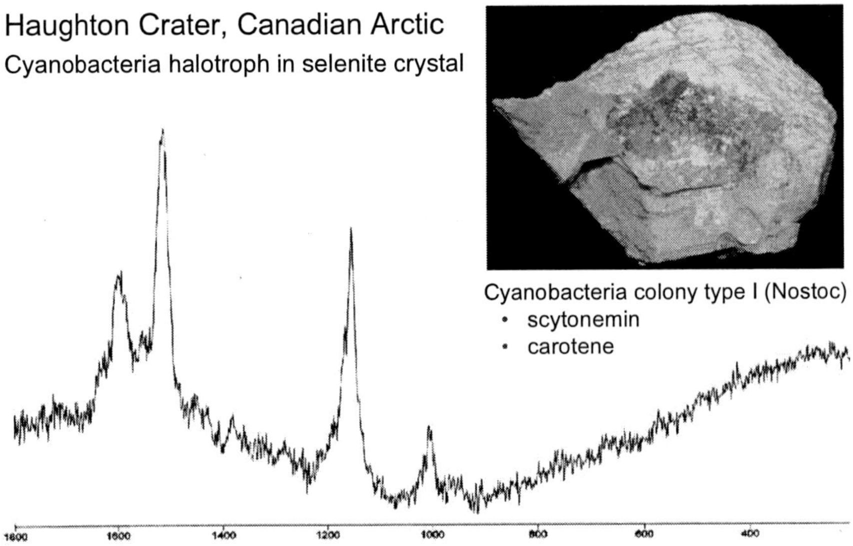

Figure 22. Raman spectra of halotrophic cyanobacterial colony inside selenite crystal, located approximately 5-8 mm below the upper surface. The Raman spectrum of the cyanobacterial inclusion obtained using confocal laser imaging shows the presence of scytonemin and a carotenoid.

A parallel study of an Antarctic gypsum endolith shows evidence of the same radiation protectant, scytonemin, in the lower spectrum with that of the surface gypsum for comparison in the upper spectrum; the different strategies adopted for the chemical protection of the colonies against low wavelength radiation for the two Antarctic endolithic systems, involving in one case the presence of haematite at the surface and in the other gypsum, is noteworthy.

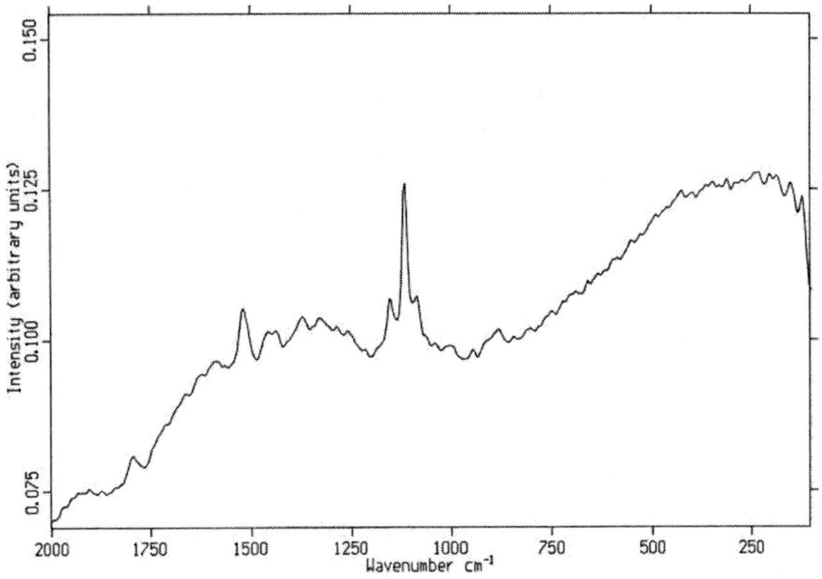

Figure 23a. Salda Golu Lake, Turkey : hydromagnesite colonised stromatolites in saltern.The FT-Raman spectrum of the cyanobacterial colonised zone in a stromatolite is shown.

Figure 23b. Salda Golu Lake, Turkey : hydromagnesite colonised stromatolites in saltern.The FT-Raman spectrum of the cyanobacterial colonised zone in a stromatolite is shown.

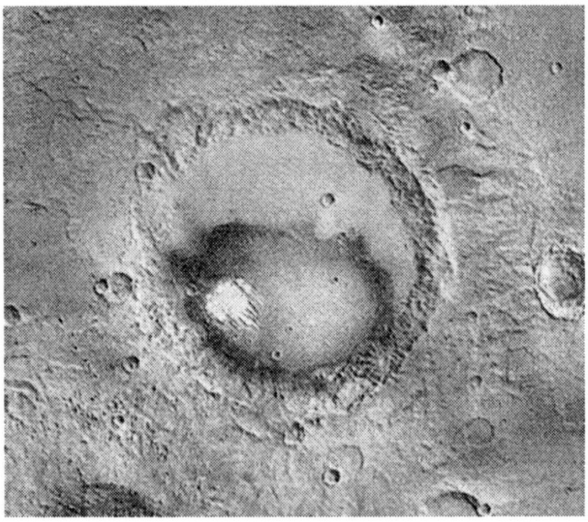

Figure 24. Juventus Chasmae, potential stromatolite region, on Mars. Reproduced from the Journal of the Geological Society, 156, Michael J. Russell *et al.*, Search for signs of ancient life on Mars : expectations from hydromagnesite microbialites, Salda Lake, Turkey, 1999, with permission from the Geological Society.

Stromatolites

Shade adaptation is required by bottom-dwelling cyanobacteria in ice-covered lakes, whilst enabling them to avoid UV-stress. During seasonal changes in the flow of melt-water, the thick cyanobacterial mats at the bottom of ice-covered Antarctic Dry Valley lakes becomes buried in silt. However, they permeate through the mineral layer to form a new mat each season, resulting in stratified stromatolites. The translucency of the 3 m thick ice cover on Lake Vanda permits penetration of solar radiation through 70 m of water column to the bottom mats where the receipt of photosynthetically active radiation (PAR) is remarkably up to ~ 120 μmol m^{-2} s^{-1}. Even the PAR level of ~ 15 μmol m^{-2} s^{-1} measured at the bottom of Lakes Hoare and Fryxell, which have thicker ice covers (about 4.5 m), is more than enough to sustain active photosynthesis by oxygenic mats. This has resulted in a wide diversity of cyanobacterial mat communities in many Antarctic Dry Valley lakes. Analogous habitats on Mars, such as Gusev and Noachis Craters, may provide suitable sources of fossil microbes and biomolecules such as porphyrins and cyanobacterial scytonemin. Modern stromatolites are considered to be analogues of *Conophyton*, which is a columnar stromatolite abundant in Precambrian rock when UV stress would have been much greater.

We have studied columnar endolithic colonisation of carbonates from Salda Golu Lake in Turkey, where stromatolitic outcrops of magnesite and hydromagnesite have formed (Figure 23a/23b). Here, the identification of the biological component in the Raman spectrum of the inorganic carbonate matrix is a good terrestrial model for the survey of potential stromatolite sites on Mars – such as that identified at Juventus Chasmae in Sabaea Terra (Figure 24) near the Schiaparelli Crater; in the figure, the light-coloured area in the meteorite impact crater is believed to be a magnesium carbonate deposit.

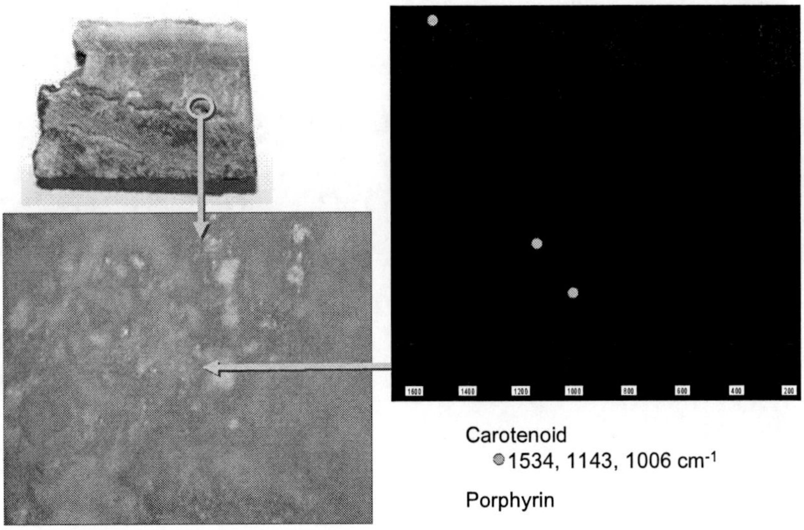

Figure 25. Trendall Pilbara stromatolite, Western Australia, 3800 Mya ; Raman spectroscopic evidence for protective biochemicals, carotene and porphyrin, produced in a protected niche

With climatic changes on a geological scale, these lakes may dry up and their cyanobacteria become desiccated fossils. This process may have helped to preserve cyanobacteria-like organisms in chert (flint) from the Australian Apex formation in Queensland for 2.5 Gya. Their pigments, especially early porphyrins which are recalcitrant components of chlorophylls, are recognizable in oil-bearing shales. A primitive photosynthetic system capable of delivering an electric charge across a membrane bi-layer could have consisted of a porphyrin (pigment), a quinone (electron donor) and a carotene (electron acceptor) to provide a chain of conjugated bonds. Such molecules are valuable biomarkers on Earth and may fulfil a similar function for their recognition by remote life-detection systems on Mars. A recently recorded microRaman spectrum of the Trendall Pilbara stromatolite has identified a carotene and a porphyrin together in a fissure inside the rock (Figure 25). This stromatolite is one of the oldest rocks found on Earth, with an age of 3800 M years, dating from a very early period in the geological evolutionary history of our planet. However, whilst it would be facile to assume that the bioorganic signatures that have been detected here are also of this age, this example does demonstrate very nicely the viability of Raman spectroscopy for the detection of organic signatures in ancient rocks. In another sampling region of the same specimen, there is clear spectroscopic evidence for the presence of beta-carotene, scytonemin, dolomite and goethite (Figure 26).

Salt Pan Evaporites

The dried terrestrial salt pans with their extremophilic halotrophic cyanobacterial colonisation are good models for the study of Martian sites such as Gusev Crater. Here, an example from the Rhub-al-Khalil in the Arabian Desert shows a cyanobacterial colony in a dolomitic zone, several cm below the surface of a large gypsum crystal embedded in a halite matrix (Figure 27). The Raman spectra of the biological component clearly demonstrate the presence of photoprotective pigments such as scytonemin and carotenoids.

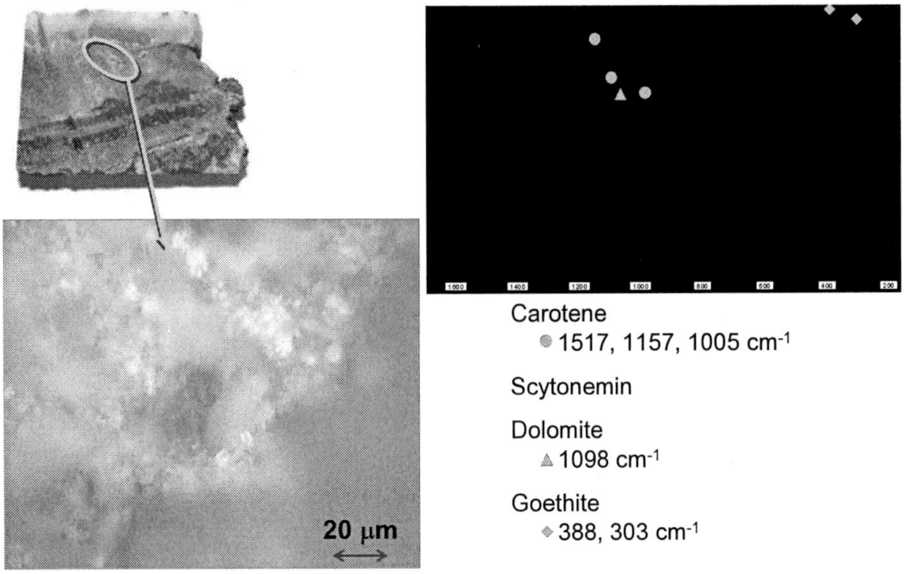

Figure 26. Specimen of Trendall Pilbara stromatolite shown in Figure 15, but different niche; different chemical suite identified –scytonemin, carotene, dolomite and goethite.

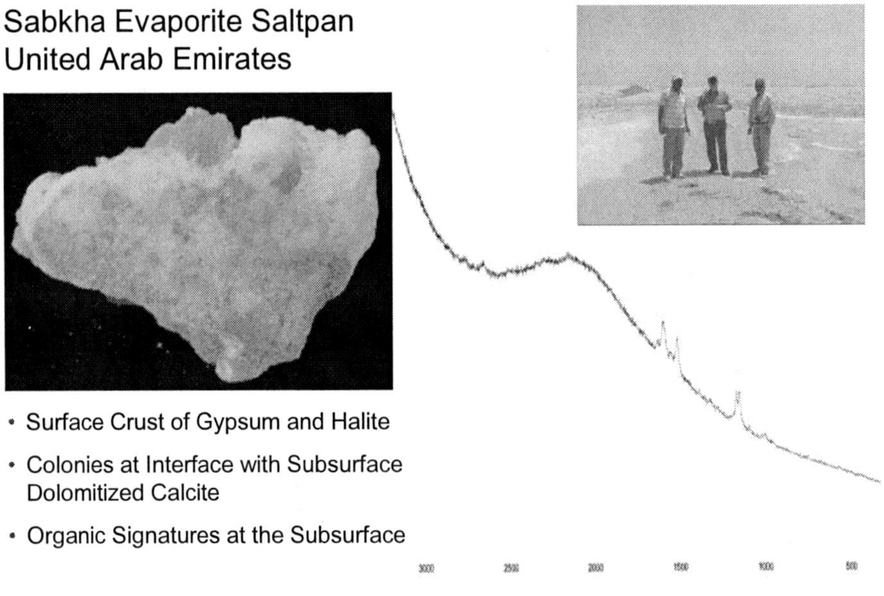

Figure 27. *Sabkha* surface crust with gypsum and halite crystals, Rhub-al-Khalil, Arabian desert ; cyanobacterial colonisation at interface with subsurface dolomitized calcite. The Raman spectrum shows the presence of scytonemin and carotene.in the biological zone.

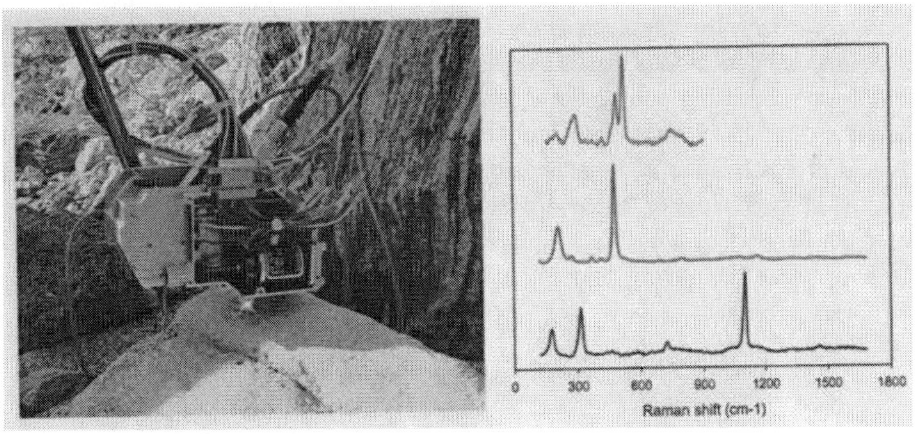

Figure 28. Prototype advanced Raman field instrument constructed for NASA adoption by JPL/Washington University St Louis ; the laser excitation is at 532 nm and the spectra from the top are of feldspar, quartz and calcite.

REMOTE SENSING PROBES

Difficulties experienced in the enclosure of a specimen in the sample illuminator of a laboratory – based Raman spectrometer has led to the evolution of remote sensing probes which bring the laser beam onto the specimen and collect the scattered Raman radiation. Typically, these probes are made of fused silica and are comprised of a single optical fibre which conducts the laser onto the specimen through an objective lens surrounded concentrically by six similar fibres which collect the scattered radiation and image this into the spectrometer- this is described as a 6/1 probe.

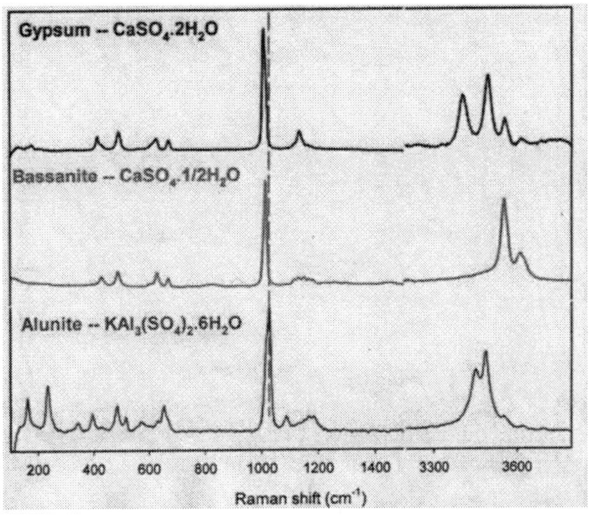

Figure 29. Raman spectra of sulfates recorded using the advanced prototype instrument shown in Figure 28 ; top, gypsum. middle, bassanite, lower, alunite.

This arrangement for remote sensing is seen to be the method of choice for the remote sensing Raman analytical experiments envisaged for extraterrestrial exploration. A problem with remote sensing using optical fibres is the superposition of the background spectrum of the silica fibre upon the Raman spectrum; an example of a remote Raman spectrum recorded with the aid of a 6/1 probe of 1 metre length is shown in Figure 28. Here, the broad bands from the Si-O stretching of silica are seen below 1000 cm^{-1}; these could obscure the mineralogical features of a specimen, but spectral subtraction of the probe background can provide a means of removing this intereference, albeit with the production of increased spectral noise.

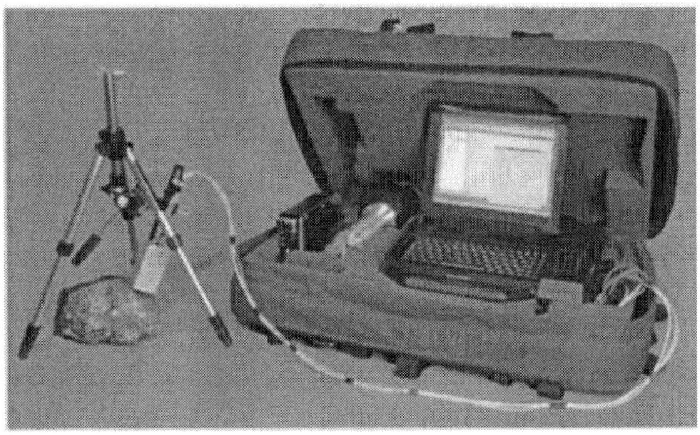

Figure 30. Prototype Raman field instrument constructed for NASA by the University of Montana; operating with laser excitation of 852 nm.

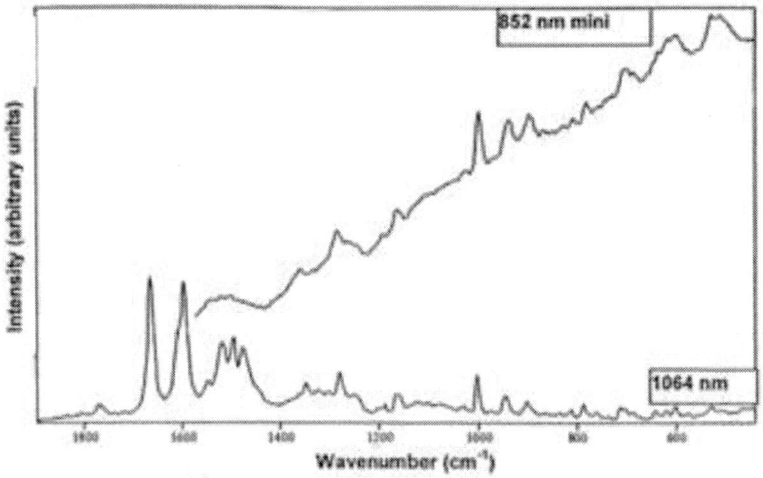

Figure 31. Raman spectra of *Acarospora chlorophana* recorded using the miniature Raman instrument shown in Figure 30; despite the limited operational range of this instrument compared with the laboratory-based FT instrument using 1064 nm laser excitation , the characteristic features of the biomolecules are clearly observed.

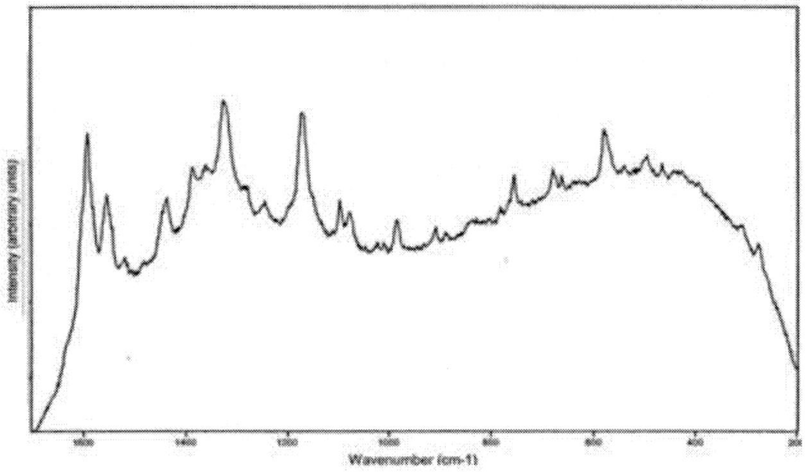

Figure 32. FT-Raman spectrum of scytonemin, the UV-radiation protectant biomolecule synthesised by several extremophilic species in hot and cold deserts

MINIATURISED RAMAN INSTRUMENTATION

Several miniaturised prototype Raman spectrometers have been constructed and are being developed and evaluated by groups for possible space mission adoption by NASA and ESA. The ExoMars/PASTEUR and AURORA ptojects are currently targeting the inclusion of novel miniaturised Raman spectroscopic instrumentation in mineralogical and life-detection suites on robotic planetary landers and rovers for the exploration of Mars.

Figure 28 shows a prototype advanced field instrument for mobile remote Raman spectroscopy from JPL/University of Washington St. Louis ; this compact unit delivers green radiation at 532 nm (frequency-doubled Nd^{3+}/YAG laser radiation) onto a specimen through the probe head of a sensing device and the results from three geological samples are shown. The upper spectrum is that of feldspar, the middle spectrum is that of quartz and the lower spectrum is calcite. The quality of the data produced by this miniature spectrometer can be gauged from the high level of discrimination evidenced in Figure 30 from the three mineral sulfates, gypsum, bassanite and alunite (results reproduced courtesy of Dr. Alian Wang [49], Washington St. Louis University). Another miniature Raman instrument that has ben constructed as a joint venture between NASA and the University of Montana [50] (Professor David Dickensheets) is shown in Figures 30 and 31. Here, 852 nm radiation in the near infrared region is used to excite the Raman spectra of organic molecules; Figure 33 shows the Raman spectrum of *Acarospora chlorophana,* in which several key features are indicated over the wavenumber range 500-1500 cm^{-1} with an operational spectral resolution of 12 cm^{-1}. This arrangement is seen not to compromise the retrieval of quality spectral data in the lower wavenumber region in comparison with the spectra that were excited at 1064 nm with the laboratory-based FT-Raman system on the same sample. However, the decreased response and sensitivity of the miniaturised instrument in the higher wavenumber region is noticeable. A comparison of the bands recorded using this miniature Raman spectrometer operating at 852 nm and laboratory instruments using other lasing wavelengths of excitation is given in

Table 4; this confirms that the identification of the protective biochemicals produced by the extremophile lichen from our standards in the database can be accomplished successfully using 852 nm radiation.

The importance of the evaluation of miniaturised Raman prototypes using terrestrial Mars analogues, particularly of extremophilic biological organisms, is stressed so that the advantages and limitations of parameters such as laser excitation wavelength, spectral wavenumber range, spectral resolution, scan characteristics and the speed of data acquisition can be properly quantified in the design of future experimental spectrometers for space mission adoption.

A detailed knowledge of the different chemical strategies adopted by terrestrial extremophiles, which are considered Martian analogues, is a prerequisite for planetary exploration. Geo – and biomarkers must be studied together to provide a better understanding of the adaptation abilities of the organisms. Geo- and bio-strategies are complementary and the survival of the organisms depends critically upon their being able to modify their microniche environments.

Table 6.

Excitation Wavelength /nm					Assignment
1064	*852/ mini*	*785*	*633*	*514.5*	•
711 w	-	713 w	713 w	-	
702 mw	701 m	703 w	703 w	-	CCH aromatic def
687 mw	687 w	687 w	687 w	-	CCH aromatic def
640 w	636 w	639 w	641 w	-	CCH aromatic def
619 w	616 w	619 w	618 w	-	CCO ring bend
599 w	600 mw	601 w	603 w	-	CCO ring bend
527 mw	529 m	529 w	529 w	-	CCO bend
508 mw	510 m	-	-	-	Calcium oxalate , bend
464 w	464 vw	-	-	-	Si-O , quartz , substrate
449 vw	-	447 w	450 w	-	
406 w	406 mw	-	-	-	Haematite , substrate
206 w	-	-	-	-	Si-O , quartz , substrate

Table 7. Raman bands of the most common and geo-and biomarkers in extremophile examplars, and their chemical formulae

Calcite	$CaCO_3$	**1086**	**712**	**282**	156								
Aragonite	$CaCO_3$	**1086**	**704**	**208**	154								
Dolomite	$CaMg(CO_3)_2$	**1098**	**725**	**300**	177								
Magnesite	$MgCO_3$	**1094**	**738**	**330**	**213**	119							
Hydromagnesite	$Mg_5(CO_3)_4(OH)_2\ 4H_2O$	**1119**	**728**	326	**232**	202	184	147					
Gypsum	$CaSO_4.\ 2H_2O$	**1133**	**1007**		669	628	**492**	**413**					
Anhydrite	$CaSO_4$	**1015**	674	628	500	416							
Quartz	SiO_2	1081	1064	808	796	696	500	542	**463**	354	263	206	128
Haematite	Fe_2O_3	**610**	500	**411**	**293**	245	226						
Limonite	$FeO(OH)nH_2O$	693	555	481	**393**	**299**	**203**						
Apatite	$Ca_5(PO_4)_3(F,CI,OH)$	1034	**963**	**586**	**428**								
Weddellite	$Ca(C_2O_4)2H_2O$	1630	**1475** **1411**	**910**	869	597	**506**	**188**					
Whewellite	$Ca(C_2O_4)H_2O$	1629	**1490**	**1463**	1396	942	**896**	865	596	521 **504**	223 207	**185**	**141**
Chlorophyll	$C_{55}H_{72}O_5N_4Mg$	1438	1387	**1326**	1287	1067	1048	**988**	**916**	**744**	**517**	351	
c-phycocyanin	$C_{36}H_{38}O_6N_4$	1655	**1638**	1582	1463	**1369**	1338	**1272**	1241	1109	1054	**815 665**	499
Beta-carotene	$C_{40}H_{56}$	**1515**	**1155**	**1006**									
Rhizocarpic acid	$C_{26}H_{23}O_6$	**1665**	1610	**1595**	**1518**	**1496**	1477	1347	1303	1002	944	902 768 448	
Scytonemin	$C_{36}H_{20}N_2O_4$	1605	**1590**	**1549**	1444	**1323**	1283	1245	**1172**	1163 984 752 675 574 270			
Calycin	$C_{18}H_{10}O_5$	1653	**1635**	**1611**	**1595**	**1380**	1344	1240	1155	1034	**960**	878 498 484	

Table 7. (Continued)

Paretin	$C_{16}H_{12}O_5$	**1671**	1631	1613	**1153**	1387	1370	**1277**	1255	**926**	571	519 467 **458**	
Usnic acid	$C_{18}H_{16}O_7$	**1694**	1627	**1607**	**1322**	**1289**	1192	1119	**992**	959	846	602	540
Emodin	$C_{15}H_{10}O_5$	**1659**	1607	1577	1557	**1298**	**1281**	942	**565**	**467**			
Atranorin	$C_{19}H_{18}O_8$	**1666**	**1658**	1632	**1303**	**1294**	1266	**588**					
Pulvinic dilactone	$C_{18}H_{10}O_4$	**1672**	**1603**	1455	**1405**	1311	**981**	**504**					
Gyrophoric acid	$C_{24}H_{20}O_{10}$	**1662**	1628	1612	1334	1304	**1291**	**1235**	**1138**	**561**			

Corroborative bands appear in bold

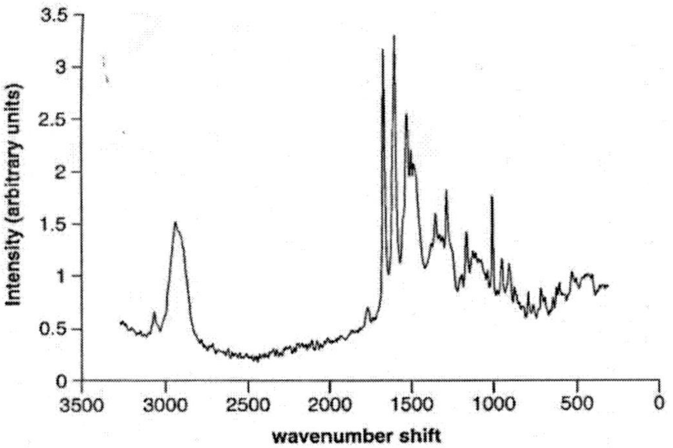

Figure 33. Raman spectrum of *Acarospora chlorophana*, an epilithic extremophile from Antarctica, recorded with the miniature Raman spectrometer shown in Figure 30 using 1064 nm laser excitation.

CONCLUSIONS

Raman spectroscopy is a useful analytical technique for planetary exploration because it does not require sample manipulation, and macroscopic and microscopic analyses are possible when identifying components in the specimen. It is sensitive to organic and inorganic compounds and is not destructive to samples that are strictly limited in quantity or accessibility and so they can be used afterwards for other, perhaps more destructive, analyses. In this respect the situation of a Raman spectrometer on a planetary rover or on a fixed lander, to which specimens are brought by a rover, are both acceptable space mission scenarios.

Laser excitation wavelength is an important parameter to consider, not only because fluorescence emission can swamp the weaker Raman scattering, but also because of the different sensitivities shown towards compound identification by different excitation wavelengths. In our assessment 785 nm laser excitation gives a good balance between organic and inorganic compounds, but 514 nm excitation induces a resonance Raman effect for carotenes whose identification is reasonably straightforward, even in small concentrations, although the higher fluorescence emission from some systems can still mask several weaker Raman bands. The observation of more than two spectral biomarker bands is necessary normally for the unambiguous identification of the components of a specimen, particularly when there is a complex mixture of organics and inorganics present.

Spectral resolution is another critical parameter that can affect observed band wavenumbers due to the overlapping of close bands, which can cause erroneous identifications of the material when comparing with standard wavenumber tables and spectral databases. The technical requirements for the construction of a miniaturized Raman spectrometer suitable for extraterrestrial planetary exploration could force a rather low spectral resolution, and so new databases adapted to low resolution spectral data would then be necessary. Nevertheless, good quality spectra are now being obtained from miniaturised Raman spectrometers such as that shown in Figure 30, which weighs one kilo and operates with diode laser excitation at 1064 nm. An example of the Raman spectrum of an extremophile obtained from this instrument in our laboratories is shown in Figure 32.

The ability to identify organic molecules and minerals present in admixture in a specimen at the same time is an advantage of the Raman analytical technique in astrobiology. Extremophiles are closely related to rock substrata, and biological and geological spectral signatures are shown together in the Raman spectrum; these are distinguishable and can be unambiguously assigned, even if some appear in a relatively low proportion. A list of some of the most important biogeological Raman spectral markers is provided in Table 7 ; this can be regarded as a prototype database for the spectral recognition of indicators from remote life detection instrumentation on planetary landers and rovers.

THE FUTURE

In 2006, the European Space Agency announced that significant funding had been allocated for the forthcoming ExoMars mission on Project AURORA – the most ambitious scientific analytical programme yet devised to detect the signatures of extinct or extant life on Mars (Figure 32). A miniaturised scientific instrumental payload on a planetary rover will use a sophisticated arsenal of analytical techniques, some of which have never before been used in space missions. The scientific instrumentation suite on ExoMars (Pasteur) has to fulfil three specific functions, namely, the characterisation of the geological context of the selected landing site, the detection of life signals and organic molecules in the samples from that site and the evaluation of the hazards to the planned human exploration of Mars. An integral part of this instrumentation will be a novel combined miniaturised Raman/LIBS spectrometer which will be designated the specific task of searching for biogeological marker signatures in the Martian surface and subsurface regolith (Figure 33). The Raman spectroscopic data presented in this paper from terrestrial Mars analogues and models will form an important part of the analytical scientific preparation for Mars exploration involving the search for life in the ExoMars mission.

The ExoMars mission and others that follow it, including the manned lunar landings, will culminate in a knowledge base that is essential for the planned human landings and exploration of Mars scheduled for 2033 – the first time that human beings will have set foot on another planet in our Solar System. A symbolism of this leap forward in analytical scientific astrobiology is depicted in Figure 36.

The authors record their appreciation of funding over many years for the basic analytical Raman spectroscopic studies of terrestrial extremophiles and Mars models from the European Space Agency and UK organisations (EPSRC, NERC and PPARC) which have enabled the preliminary work to be carried out to establish the viability of the Raman technique for incorporation into extraterrestrial life-detection instrumentation and for the opportunity to take novel analytical chemical science to the "last frontier".

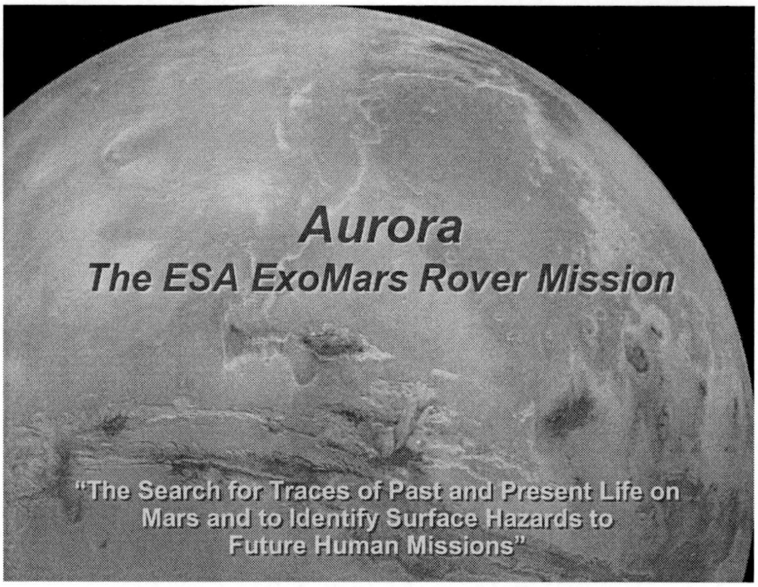

Figure 34. AURORA: the European Space Agency /NASA programme, incuding the ESA ExoMars mission.

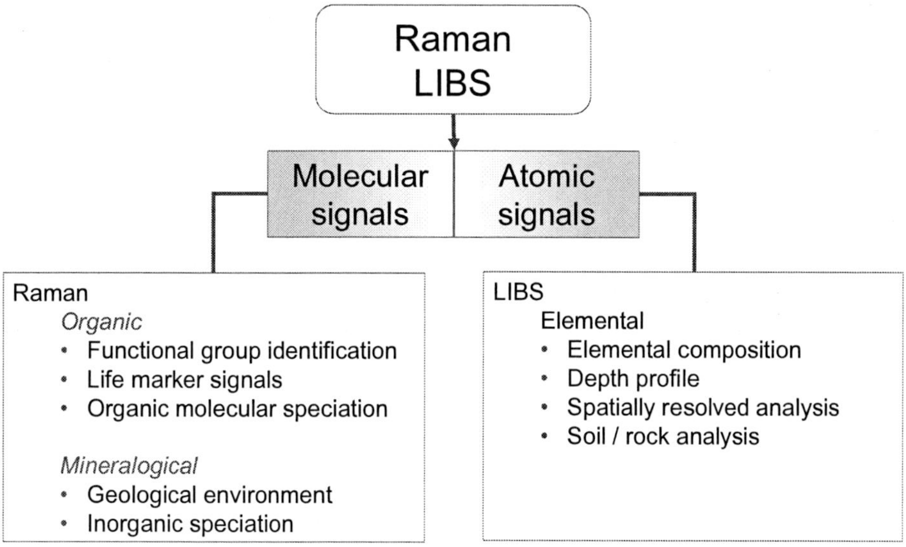

Figure 35. The principle of the novel miniaturised Raman/LIBS combined spectrometer for the ExoMars mission to Mars.

Figure 36. Symbolic depiction of the Mars mission phases, 2007-2033, involving a revisitation of manned lunar exploration before embarking upon the manned missions to Mars.

In this they are especially grateful for scientific collaboration with the late Dr David Wynn-Williams, Head of Antarctic Astrobiology at the British Antarctic Survey, Cambridge, and for the ongoing support of Professors Charles Cockell, John Parnell, Monica Grady, Fernando Rull and Drs Cynan Ellis- Evans, Roger Worland, Andre Brack, Frances Westall, Susana Jorge, Derek Pullan, Beda Hoffmann, Jan Jehlicka and Liane Benning—a range of expertise encompassing geology, microbiology, geochemistry, meteoritics, earth sciences and biomolecular sciences that is so critical for the effective study of *astrobiology,* the quest for life in the Solar System and beyond.

REFERENCES

[1] IR. Lewis, HGM. Edwards, (Eds.), Handbook of Raman Spectroscopy : *From Research Laboratory to the Process Line*, M. Dekker, New York, 2000.

[2] DA. Long, Raman Spectroscopy, John Wiley and Sons, *Chichester*, UK, 2002.

[3] ftp://ftp.estec.esa.nl/pub/aurpra/Rover/PPP-IDD.pdf ; *ESA Progress Letter*, (2004) : Pasteur Instrument Payload for the ExoMars Rover Mission. ESA Publications Division, Noordwijk, The Netherlands; http://esamultimedia.esa.int/docs/Aurora/ Pasteur_ Newsletter_4.pdf.

[4] Lin-Vien, D; Colthup, NB; Fateley, WG; Grasselli, JG. *The Handbook of Infrared and Raman Characteristic Frequencies of Organic Molecules*, Academic press, London, 1991.
[5] Bouchard, M; Smith, DC. *Spectrochimica Acta*, Part A, 2003, 59, 2247-2266.
[6] G; Dent, WE. Smith, (Eds.), *Modern Raman Spectroscopy*, John Wiley and Sons, Chichester, UK, 2005, 128-129 in ref. 6.
[7] Edwards, HGM. "Spectra –Structure Correlations in Raman Spectroscopy", in Handbook of Vibrational Spectroscopy, Volume 3, JM; Chalmers, PR. Griffiths, (Eds). John Wiley and Sons, Chichester, UK (2000).
[8] Edwards, HGM; Carter, EA. "Raman Spectroscopy of Biological Molecules", in *Infrared and Raman Spectroscopy of biological Materials*, HU; Gremlich, B; Yan, Dekker, M. (Eds). New York, 2001.
[9] Veronelli, M; Zerbi, G; Stradi, R. *J. Raman Spectroscopy*, 1995, 26, 683-692.
[10] A; Brack, B; Fitton, F; Raulin, A. Wilson, (Eds.), *Exobiology in the Solar System and the Search for Life on Mars*, ESA Special Publication SP-1231, ESA Publications Division, Noordwijk, The Netherlands, 1999.
[11] de Faria, DLA; Silva, SV; de Oliveira, MT. *J. Raman Spectroscopy*, 1997, 28, 873-878.
[12] Wang, H; Mann, CK; Vickers, TJ. *Applied Spectroscopy*, 2002, 56, 1538-1544.
[13] Hendra, PJ. *in Handbook of Vibrational Spectroscopy*, JM; Chalmers, PR. Griffiths, (Eds). Volume 2, John Wiley and Sons, Chichester, UK (2000).
[14] Tabaksblat, R; Meier, RJ; Kip, BJ. *Applied Spectroscopy*, 1992, 46, 60-68.
[15] Wijnaendts van Resandt, RW; Marsman, HJB; Kaplan, R; Davoust, J; Steilzer, EKH. Sticker, R. *J. Microsc*, 1985, 38, 29-34.
[16] T. Wilson, (Ed.), *Confocal Microscopy*, Academic Press, London. 1994.
[17] Hadjoost, S; Yarwood, JA. *Applied Spectroscopy*, 1996, 50, 558-564.
[18] Everall, NJ. *Applied Spectroscopy*, 2000, 54, 773-782.
[19] Everall, NJ. *Applied Spectroscopy*, 2000, 54, 1515-1520.
[20] Baldwin, KJ; Batchelder, DJ. *Applied Spectroscopy*, 2001, 55, 517-524.
[21] Michielsen, S; *J. Applied Polymer Science*, 2001, 81, 1662-1669.
[22] McDonald, AM; Vaughan, AS; Wyeth, P. *Applied Spectroscopy*, 2003, 57, 1475-1481.
[23] Malin, MC; Edgett, KS. *Science*, 2000, 288, 2330.
[24] Carr, MH. "*Water on Mars*", Oxford University Press, New York, 1996.
[25] Wynn-Williams, DD; Edwards, HGM. Environmental UV Radiation:Biological Strategies for Protection and Avoidance, in "Astrobiology": The Quest for Life in the Solar System", G; Horneck, C. Baumstarck-Khan, (Eds). *Springer-Verlag*, Berlin, 2000, 244-260.
[26] Cockell, CS. *Planetary and Space Sciences*, 2000, 48, 203.
[27] Caricchidi, R. *Astrobiology*, 2002, 2, 281.
[28] Cockell, CS; Knowland, J. *Biol.Revs.*, 1999, 74, 311.
[29] Wynn-Williams, DD; Edwards Icarus, HGM. 2000, 144, 486.
[30] A; Brack, B; Fitton, F; Raulin, A.Wilson, (Eds.), "*Exobiology in the Solar System and the Search for Life on Mars*", ESA Special Publication (SP-1231), ESA Publications Division, Noordwijk, The Netherlands, 1999.
[31] Jorge Villar, SE; Edwards, HGM. *Analytical &Bioanalytical Chemistry*, 2006, 384, 100-113.
[32] IR; Lewis, HGM. Edwards, (Eds.),"*Handbook of Raman Spectroscopy: From the Process Line to the Laboratory*", Marcel Dekker, New York, 2000.
[33] Jorge Villar, SE; Edwards, HGM; Benning, LG. *Astrobiology*, submitted for publication 2006.
[34] Lane, MD; Christensen, PR. *Icarus*, 1998, 135, 528.

[35] Catling, DC; Geophys, J. *Res-Planets*, 1999, 104, 16453-16469.
[36] Bridges, JC; Catling, DC; Saxton, JM; Swindle, TD; Lyon, IC; Grady, MM. *Space Sci. Revs.*, 2001, 96, 365-392.
[37] Madden, MEE; Bodnar, RJ; Rimstidt, JD. *Nature*, 2004, 431, 821-823.
[38] Edwards, HGM; Jorge Villar, SE; Parnell, J; Cockell, CS; Lee, P. *Analyst*, 2005, 130, 917-923.
[39] Vaniman, DT; Bish, DL; Chipera, SJ. Fialips, CI; Carey, JW. Feldman, WC. *Nature*, 2004, 431, 663-665.
[40] Arvidson, RE; Poulet, F; Bibring, JP; Wolff, M; Gendrin, A; Morris, RV; Freeman, JJ; Langevin, Y; Mangold, N; Bellucci, G. *Science*, 2005, 307, 1587-1591.
[41] Langevin, F; Poulet, J; Bibring, P; Gondet, B. *Science*, 2005, 307, 1584-1586.
[42] Murray, JB; Muller, JP; Neukum, G; Werner, SC; van Gasselt, S; Hauber, E; Markiewicz, WJ; Head, JW; Fong, BW; Page, D; Mitchell, KL; Portyankina, G. *Nature*, 2005, 434, 352-356.
[43] Cabrol, NA. Grin, EA; Pollard, WH. *Icarus*, 2000, 145, 91-207.
[44] Moore, JM; Willhelms, DE. *Icarus*, 2001, 154, 258-276.
[45] Paige, DA. *Science*, 2005, 307, 1575-1576.
[46] Hiesinger, H; Head, JW. *Planetary and Space Sciences*, 2002, 50, 939-981.
[47] Ruff, SW; Christensen, PR; Clark, RN; Kieffer, HH; Malin, MC; Bandfield, JL; Jakosky, BM; Lane MD; Mellon, MT; Presley, MA. *J.Geophys.Res-Planets*, 2001, 106, 23921-23927.
[48] Edwards, HGM; Jorge Villar, SE; Parnell, J; Cockell, CS; Lee, P. *Analyst*, 2005, 130, 917-923.
[49] Wang, A; Haskin, L; Cortez, E. *Applied Spectroscopy*, 1998, 52, 477-487.
[50] Dickensheets, DL; Wynn-Williams, DD; Edwards, HGM; Crowder, C; Newton, EM. *J. Raman Spectroscopy*, 2000, 31, 633-677.

In: Astrobiology: Physical Origin, Biological Evolution... ISBN: 978-1-60741-290-8
Editors: S. Hegedus, J. Csonka, pp. 193-204

Chapter 9

EUROPA: NEW HORIZONS FOR ASTROBIOLOGY*

Esther Velasco[1], Guillermo Caravantes[2,3], Beatriz Gómez-Izquierdo[2,3], Alberto Rodriguez[3] and Javier Ruiz[1]

[1] Centro de Biología Molecular,
CSIC-Universidad Autónoma de Madrid,
28049 Cantoblanco, Madrid, Spain
[2] Volcano Dynamics Group, The Open University,
MK7 6AA, United Kingdom
[3] Seminario de Ciencias Planetarias, Facultad de Ciencias Geológicas,
Universidad Complutense de Madrid, 28040 Madrid, Spain

ABSTRACT

Europa is the smallest of the Galilean satellites of Jupiter, although it is however a remarkable body and the primary focus of astrobiological interest in the Jovian system. Geological and magnetic evidences, as well as theoretical considerations, strongly suggest that Europa has a thin ice shell, maybe a few kilometers or a few tens of kilometers thick, which floats on an internal ocean of liquid water. Some anomalies have been observed in the huge magnetic field of Jupiter whenever Europa goes through it, which seems to indicate that this ocean is made of conductive water, due to a probable high content of salts. Moreover,the low density of impact craters suggest an surface age of about 60 Myr in average, which is very young in geological terms, indicating that internal activity, driven by tidal heating, is resurfacing the moon at the present time. So, Europa shows an overprint of crisscrossing ridges and linear markings, some extending for thousand of kilometers, which are locally disrupted for chaotic terrains, building a colorful and exotic textured surface. All of these features make of Europa a theoretically good niche for life. Biological organisms might exist in the salty internal ocean, in the icy shell (active or latent) at places such as fractures, veins or liquid reservoirs, on the

* A version of this chapter was also published in Space Exploration Research, edited by John H. Denis and Paul D. Aldridge, Nova Science Publishers. It was submitted for appropriate modifications in an effort to encourage wider dissemination of research.

internal sea floor in hydrothermal vents or submarine volcanoes, or even within the rocky interior.

INTRODUCTION

Europa, the fourth largest Jovian satellite, included in the four Galilean satellites, has about the same size (and slightly lower density) as the Earth's Moon, but its surface is between the brightest in the Solar System as consequence of sunlight reflecting from an icy shell. Talking about geological structures, the surface of Europa shows a wide variety of geologic terrains and features (Lucchita and Soderblom, 1982; Malin and Pieri, 1986; Greeley et al., 1998, 2000), with local prominent relieves of up to 1 km (Schenk, 2002). From the scarcity of impact craters, an age of $\sim 10^7$ years has been suggested for the surface (Zahnle et al., 2003). However, after the formation of these craters, almost no tectonics has deformed them, what may mean a decrease in tectonic activity in a moment in a recent past.

Moreover, there are solid evidences for the existence of an internal ocean of liquid water below the icy shell of this satellite of Jupiter. The more strong evidence comes from magnetic data acquired by the Galileo spacecraft, which recorded magnetic anomalies in the huge magnetic field of Jupiter whenever Europa goes through it (Kivelson et al., 2000). These observations indicates that there is a global conductive electric layer locate ~20 km below the surface, and it could be explained by an internal ocean made of salty water (Kivelson et al., 2000; Schilling et al., 2004) maybe compositionally similar to that of the Earth's oceans. This internal ocean could be habitable (e.g., Reynolds et al. 1983; Chyba 2000; Chyba and Phillips 2001, 2002; Schulze-Makuch and Irwin 2002; Marion et al. 2003). Similarly, biological organisms may exist, active or latent, inside the icy shell above this internal ocean, at places such as fractures, veins and liquid reservoirs (e.g., Greenberg et al., 2000; Marion et al., 2003; Lipps and Rieboldt, 2005), and inside and around warm ice diapirs (Price, 2003; Ruiz et al., 2007).

In order to understand the biologic potential of Europa and facilitate the search for possible ecological niches, we must review the present knowledge of the ice shell, and its structure and chemical-physical conditions, due to its key role as a direct or indirect communication between ocean and surface, and as a potential marker of the ocean's composition.

EUROPA'S SURFACE: A COMPLEX GEOLOGIC RECORD

Europa seems to be a geologically young world, with a mean age about 60 Myr (Zahnle et al., 2003), and may therefore continue to be active today. This conclusion is based on the relative low density of impact craters observed on its surface. The surface shows, mainly, three kinds of terrains: bright plains, mottled terrains and disrupted regions (For a review of the geology of Europa see Greeley, 2004). They are (more or less) mixed on the Europa surface, which exhibits an overprint of crisscrossing ridges and linear markings, some extending for thousand of kilometers. These near linear features are locally disrupted for chaotic terrains, building a colorful and exotic textured surface.

The bright plains seem to be the basic unit of Europa's surface, and many of the rest of units can be derived, mainly, from that one. These plains show plateaus up to 10 km wide and tens of meters high with, in some places, multiple sets of ridges and grooves with a length of a few tens meters wide. This structure could be formed by repeatedly tectonic processes, which deform a relative low-density ice of the icy shell. The mottled terrains show an irregular patchwork of darker zones, of a range in size from ~50 km to ~500 km across or more, with irregular margins grading into the bright plains. This terrain could be formed in response to endogenic forces, which have broken bright plains into plates with intervening areas of chaotic terrain. This low-lying chaotic terrain is darker than disrupted plates and probably contains a greater proportion of non-ice material. So, this terrain is one of the most important places to looking for biosignatures. Other important places on the surface are the so-dubbed "lenticulae", which are round, elliptic or irregular shaped features, including dark spots, pits, domes, microchaos areas and are considered an evidence of the younger geological (endogenic) activity recorded on Europa (Pappalardo et al., 1998; Prockter et al., 1999; Figueredo and Greeley, 2000); their size goes from less than ten to several tens of kilometers, and in some areas are highly concentrated (Greeley et al., 1998; Pappalardo et al., 1998).

The differences of color and spectral properties between terrains cannot be explained in a simple way. The shell composition may include silicate minerals, magnesium and sodium sulfate hydrates, sulfuric acid, and so on (McCord et al. 1998; Carlson et al. 1999, Kargel et al. 2000; Spencer et al. 2006), and these most likely are of internal origin. The abundance of biologic elements on Europa is still unknown.

Thickness of the Icy Shell

One of the main topics regarding the existence of life is the ice shell thickness, because it determines the model and dynamics of the ice shell and therefore constraint the physical conditions in the inner ocean. Two models have been proposed to explain the observed features and physical parameters, and constitute the famous controversy thin shell vs. thick shell.

For the thin shell model (Greenberg et al., 1999, 2000; Hoppa et al., 1999) some features, such as the so-called cycloids, would seem to suggest the presence of an ocean a few kilometers below the surface, which would be locally exposed in the surface due to the melt-through of a thermally conductive ice shell. It would be warmed from below, with a surface temperature of about 100 K and a maximum thickness of a few kilometers. It would be required significant tidal heating in the rocky interior to maintain such thin shell, maybe including a molten silicate layer. On the other hand, the thick shell model proposes a shell a few tens of kilometers thick. After this model, the icy shell can be subdivided into a upper, cold and inmovil, stagnant lid, and a thick layer of ductile, maybe convective, ice (Pappalardo et al., 1999a,b; Pappalardo and Head, 2001).

The detailed analysis of size and depth of some of the major impact structures provides a thickness of ~19-25 km for the ice shell (Schenk, 2002), and the important amplitude of surface topography also supports the thick shell model (Prockter and Schenk, 2005). Moreover, the most accepted hypothesis to explain the origin of the lenticulae proposes that these features reflect the effects on the near surface crust of convective upwelling plumes ("thermal diapirs"); in this case, the lenticulae would be showing us the recent activity of the convective system, and the spacing between them would indicate that the thickness of the convective layer is ~10-20 km thick (Pappalardo et al., 1998; Spaun et al., 2004).

So, the evidences favoring a, at least relatively, thin shell, imply that the communication between the surface and the internal ocean would be greatly restricted to thermal diapirism (linked to the convection), and maybe cryovolcanism and meteoritic impacts.

Thermal State of the Icy Shell

There is scarce information about the heat flow of Europa, although some estimation have been proposed based on relation between geological features and the mechanical and thermal properties of the lithosphere. For Europa, it is usual to put the brittle-ductile transition 2 km deep at most (e.g., Pappalardo et al., 1999b). For example, the undulations found in Astypalaea Linea, interpreted as folds (Prockter and Pappalardo, 2000), have a wavelength of ~25 km, which would imply a brittle-ductile transition ~2 km deep (Dombard and McKinnon, 2006); undulations with a similar spacing have also been observed in the leading hemisphere of Europa (Figueredo and Greeley, 2000). On the other hand, troughs, possibly grabens, in the Callanish and Tyre multiring impact structures are up to ~2 km wide: if these troughs are interpreted as grabens then its width implies a depth of faulting of ~2 km, which could correspond to the brittle-ductile transition (Ruiz, 2005; Lichtenberg et al., 2006). For this depth the vertical heat flows must be close to 100 mWm^{-2} (Ruiz and Tejero, 2000; Ruiz, 2005; Dombard and McKinnon, 2006; Lichtenberg et al., 2006).

On the other hand, the effective elastic thickness of the lithosphere varies depending on the region. Available estimations range from ~0.4 below ridges and domes to ~2.9 km below a plateau in the crater Cilix region. The lower values correspond to geological setting where tidal heating can be differentially enhanced, and here not representative for average Europa. For the Cilix region the estimated heat flow would be ~25-35mWm^{-2} (Nimmo et al., 2003; Ruiz, 2005), value lower than those deduced from the depth brittle-ductile transition, which suggest the existence of local variations in the thermal state of the ice shell, maybe related to variations in the vigor of the convection (Ruiz et al., 2005).

These indications obtain heat flow high values when compared to the component produced in the rocky fraction by radiogenic heat at the present, which is ~6-8 mWm^{-2} (Cassen et al., 1982; Spohn and Schubert, 2002). This implies an important role for tidal heating, what should happen mainly in the warm inner part of a convective layer (McKinnon, 1999; Ruiz and Tejero, 2003). Indeed, some theoretical studies of tidally heated convection in the ice shell of Europa obtain heat flows of ~100 mW mWm^{-2} (Nimmo and Manga, 2002; Ruiz and Tejero, 2003; Moore, 2006). Because convection could start in the shell for heat flows under 10-45 mW m^{-2} (the exact value depends on the particular model; e.g., Ruiz and Tejero, 2003), most of the surface heat flow would be the result of tidal heating in the convective layer. On the other hand, a conductive ice shell at least ~19-25 km thick (Schenk 2002) implies that the surface heat flow should be at most ~20-30 mW m^{-2} (Ruiz, et al., 2005). This implies a heat flow of at most ~30-45 mW m^{-2} reaching the shell base (or equivalently, the sea floor), which limit the possibility of hydrothermal circulation in the rocky portion of Europa.

LIMITING FACTORS FOR LIFE IN EUROPA

We have not found any definitive probe of life outside the Earth. However, on the Earth, life can resist conditions that we thought life would not be able to resist. Also, the conditions that we think they are optimal for life (293 K, pH = 7, and so on) are an environment where some kinds of life are not able to live. In fact, Europa has some places where life (as we know life), would have able to take place and develop in a different way with respect to Earth. So, a review the limiting factors for Earth's life is necessary for an assessment of the possibility of life on Europa.

Temperature

On Earth, temperature tolerate for life usually going to -20° to 121° C (Marion et al., 2003). However, recently has been discussed evidence for microbial metabolic activity in Antarctic glacial ice at -40°C (Price and Sowers, 2004). It is important to know why the temperature is a limiting for life. 100° C is the temperature of the normal boiling point of water, but in hydrothermal vents and the subterranean deep biosphere, the high pressures impede water to boil, because the water molecules are kept sufficiently together for the chemical bonds are joining enough to keep the liquid state. So, in these cases, life can resist up to 110-121°C.

There are several problems for life under the freezing point. Firstly micro-crystals of ice are formed inside the cells, which broke cellular structures, not only for the effect of the micro-crystals themselves, but because water in solid state has more volume than water in liquid state. Secondly, cells need water movement through its membrane to equilibrate the ion's internal concentration; so, life cannot use icy water and have dehydration problems even in presence of abundant water in form of ice. To avoid this problem, Earth's life could have adopted strategies to reduce the water freezing point, both in the extra- and intra-cellular environments, by using cryoprotectant substances, for example through the accumulation of compatible solutes with the intracellular metabolism, as trehalose, glycerol or glycine betaine, substances that are efficient cryoprotectants (e.g., Welsh, 2000; Yancey, 2005). Moreover, terrestrial life at low temperature is adapted to low water activity in the same way as it is adapted to high osmolarity or desiccation conditions (e.g., Potts, 1994; Thomas and Dieckmann, 2002), implying that a decrease in temperature would be equivalent to an increase in osmolarity or desiccation. For this reason, it has been suggested (Ruiz et al., 2007) that the synthesis of compatible solutes can be universal metabolic strategy for hypersaline, desiccation or low temperature conditions, and for any organism living within the ice shell of Europa.

Moreover, life could be possible as long as liquid water is available. So, the term of "eutectophiles" has been used (Deming, 2002) for these hypothetical organisms living at liquid veins or reservoirs close to the eutectic point of water containing impurities.

This increases the potential habitats for life inside the icy shell of Europa.

Salinity

On Earth, we can find a lot of kind of hypersaline habitats (Madigan, 1997; Marion, 2003). Microorganisms adapted to high salinity are called halophiles, or even extreme halophiles organisms, when, not only can resist high salinity, but also, they need a high salinity to be alive. These extreme halophiles, mainly inside the group of unicellular microorganisms called archaeas, can have metabolic activity in salinity close to saturation (Madigan, 1997). So, the briny ocean of Europa could be a habitable one.

Acidity

On Earth, exist examples of extreme pH tolerances (e.g., Bachofen, 1986; Zhilina and Zarwarzin, 1994; Johnson, 1998; Scheleper et al., 1995; Marion 2003), from < 1 to > 11. For this reason, the pH value seems not a great limiting factor for Earth's life, and all the proposed compositions for the Europa's ocean (Kargel et al., 2000; Marion, 2001, 2002) could, in theory, support life.

Radiation

Although there are organism on Earth that could survive under ultraviolet and ionizing radiation (e.g., Bacholen 1986; Dose et al 2001; Rothschild, 2001), which could be the great pitfall for the possibility of Europan life near surface. The resistance to this kind of radiations lies in the capacity of the organism to repair DNA quickly and not to an immune skill of the leaving beings.

Pressure

Some authors think that pressure, or the relation between temperature and pressure, is a limiting factor for organisms on Earth. It could be true in giant gas planet, where the high temperature exceed amply the known range of temperature for life, or underground, some kilometers of dip inside the crust. Nevertheless, in aquatic environment, pressure are not a problem for life in general (some organism could be very sensitive to changes of pressure, others not, but it is not a problem for life), because the water has the next property, the heaviest or the most dense water has a temperature of ~4° C. Warmer water floats to upper parts in the water column, and colder water and ice floats on the surface. So, 4° C is a temperature inside the range of life temperature, and there are lots of organisms which can live in that temperature. And only thinking in pressure effects, it could be a problem for life if there were high variations of pressure in aerial, or partially aerial, ecosystems. However, in aquatic systems, it would not have to be a problem, because water has a constant volume with pressure variations, and the pressure is the same outside and inside the cell or organism. Then, all the ocean's floor could be a possibly habitat for life.

WHERE LOOK FOR?

Europa is a world with a special fascination and interest that may help us to understand very aspect hat we don't know yet, for example: what are the limits inside which Life, such as we know and understand it, can exist? Is it possible that Life is a more frequent process than we thought? Earth contains the unique biosphere in the Solar system?

There are different ecosystems considered to be potentially viable for support life on Europa. Chyba and Hand (2001) and Lipps and Rieboldt (2005), have correctly argued that a photosynthetically-driven ecosystem is not impossible, but doubtful. Photosynthesis could be possible close to the surface, for organisms living at liquid veins or reservoirs inside the icy shell. But sunlight is not as accessible as on the Earth, and for this reason it is more probable the existence of ecosystems based in other modes of energy supply. Zolotov and Shock (2004), have proposed several possible biogeochemistry models for Europa, mainly based on sulfur, carbon and iron reduced compounds.

On Earth, the best studied non-photosynthetic ecosystems are marine hydrothermal vents, where chemoautotrophic organisms use mainly the metabolism of H_2S and FeS to obtain nourishment through the oxidation of these inorganic chemical compounds. So, the internal sea floor, on the possible hydrothermal vents or submarine volcanoes, could be a good candidate for an Europan niche, although it is not clear if hydrothermal activity exists. Other possibilities for habitats are the Europa's rocky interior (similarly to Earth underground ecosystems), the water column and the water-icy shell interface. The icy shell can harbor life, inside of liquid veins or reservoirs, or close to thermal diapirs (Ruiz et al., 2007). Even possible habitats on the Europa's surface have been described (Lipps and Reiboldt, 2005), but the superficial conditions are really extreme in that habitats, at least for life as we know.

CONCLUSION

Although several possible niches which could potentially maintain Europan life, this life would be subject to extreme conditions, at least on the basis of what we considerate extreme conditions on our planet. But, life can exist on Earth under most of these conditions.

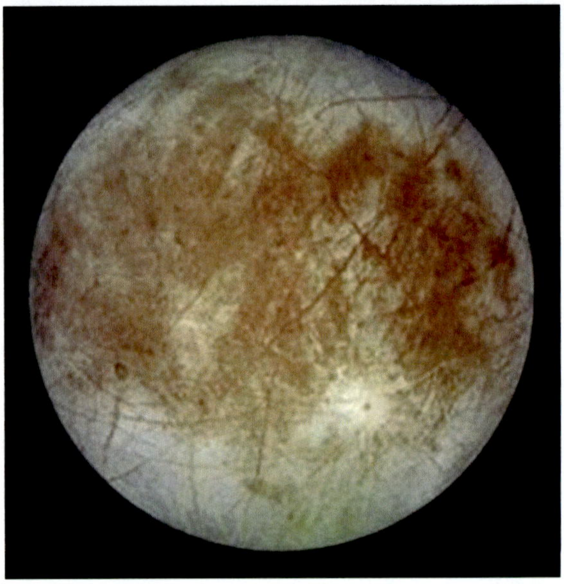

Figure 1. Global view of Europa. This image reveal the three basic elements of the geology of Europa: bright plains, mottled terrain, and linear features.

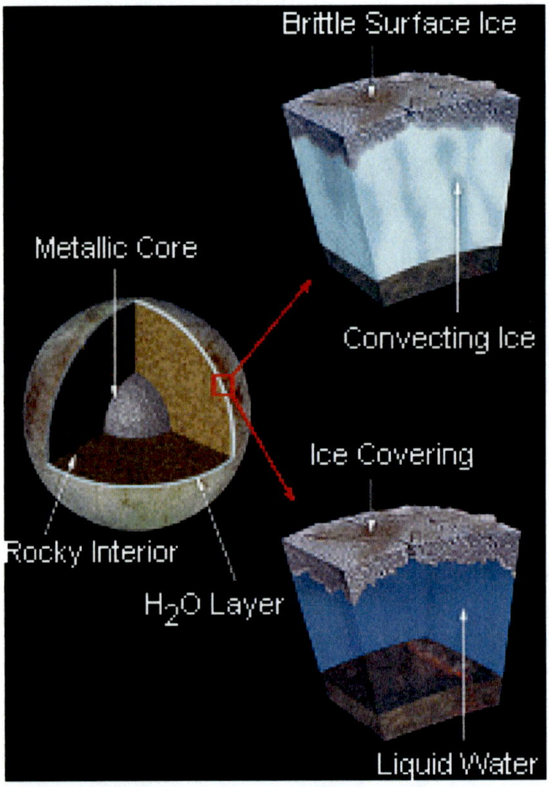

Figure 2. Europa's Interior Models: the upper one considers that there is a convecting ice layer under a brittle ice cover, and the lower one considers that there is a liquid ocean under the ice covering. The second one is supported by magnetic anomalies data (source: www.jpl.nasa.gov).

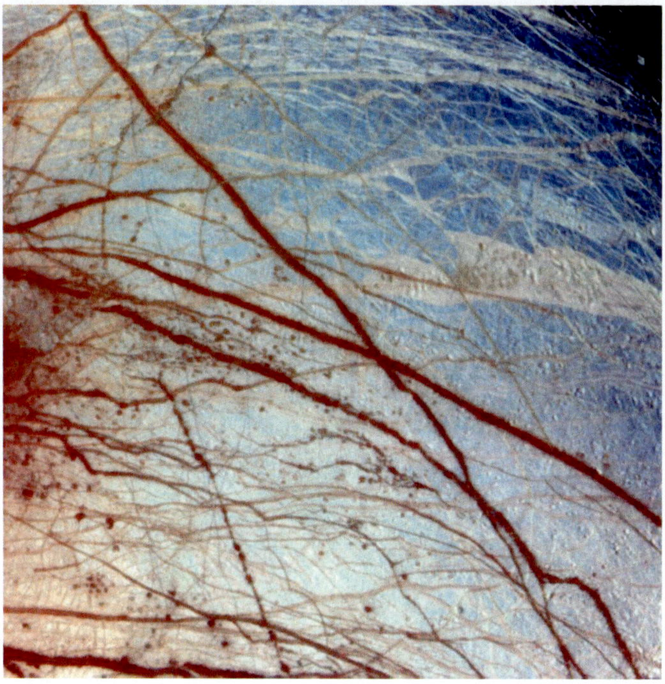

Figure 3. False-color composite of visible and infrared images from the Galileo spacecraft. The image is centered at 45° N- 221° W and shown an area 1,260 km across.

REFERENCES

Bachofen, R. 1986. Microorganisms in extreme environments: introduction. *Experientia.* 42, 1179–1182.

Carlson, R. W., Johnson, R. E. and Anderson, M. S.: 1999, Sulfuric acid on Europa and the radiolytic sulfur cycle, *Science.* 286, 97-99.

Cassen, P.M., Peale, S.J., Reynolds, R.T., 1982. Structure and thermal evolution of the Galilean satellites. In: Morrison, D. (Ed.), Satellites of Jupiter. Univ. of Arizona Press, Tucson, pp. 93–128.

Chyba, C.F., 2000. Energy for microbial life on Europa. *Nature.* 403, 381-382.

Chyba, C.F., Phillips, C.B., 2001. Possible ecosystems and the search for life on Europa. *Proc. Nat. Acad. Sci.* 98, 801-804.

Chyba, C.F., Phillips, C.B., 2002. Europa as an above of life. Origin Life Evol. *Biosphere.* 32, 47-68.

Deming, J.W., 2002. Psychrophiles and Polar regions. *Curr. Opin. Microbiol.* 5, 301-309.

Dombard, A.J., McKinnon, W.B., 2006. Folding of Europa's icy lithosphere: An analysis of viscous-plastic buckling and subsequent topographic relaxation. *J. Struct. Geol.* 28, 2259–2269.

Dose, K., Bieger-Dose, A., Ernst, B., Feister, U., Gomez-Silva, B., Klein, A., Risi, S., Stridde, C.,2001. Survival of microorganisms under the extreme conditions of the Atacama Desert. *Orig. Life Evol. Biosph.* 31, 287–303.

Figueredo, P.H., Greeley, R., 2000. Geologic mapping of the northern hemisphere of Europa from Galileo solid-state imaging data. *J. Geophys. Res.* 105, 22629-22646.

Greeley, R., et al., 1998. Europa: initial Galileo geological observations. Icarus 135, 4–24.

Greeley, R., Chyba, C.F., Head, J.W., McCord, T.B., McKinnon, W.B., Pappalardo, R.T., Figueredo, P.H., 2004. Geology of Europa, in Bagenal, F., et al. (Eds.), *Jupiter: The Planet, Satellites and Magnetosphere,* Cambridge Univ. Press, Cambridge, pp. 329-362.

Greenberg, R., Hoppa, G.V., Tufts, B.R., Geissler, P., Riley, J., Kadel, S., 1999. Chaos on Europa. *Icarus.* 141, 263-286.

Greenberg, R., Geissler, P., Tufts, B.R., Hoppa, G.V., 2000. Habitability of Europa's crust: The role of tidal-tectonic processes. *J. Geophys. Res.* 105, 17551-17562.

Hoppa, G.V., Tufts, B.R., Greenberg, R., Geissler, P.E., 1999. Formation of cycloidal features on Europa. *Science.* 285, 1899-1902.

Johnson, D.B., 1998. Biodiversity and ecology of acidophilic microorganisms. *FEMS Microbiol. Ecol.* 27, 307–317.

Kargel, J.S., Kaye, J.Z., Head, J.W., Marion, G.M., Sassen, R., Crowley, J.K., Ballesteros, O.P., Grant, S.A., Hogenboom, D.L., 2000. Europa's crust and ocean: Origin, composition, and the prospects for life. *Icarus.* 148, 226-265.

Kivelson, M.G., Khurana, K.K., Russell, C.T., Volwerk, M., Walker, R.J., Zimmer, C., 2000. Galileo magnetometer measurements: a stronger case for a subsurface ocean at Europa. *Science.* 289, 1340-1343.

Lichtenberg, K.A., McKinnon, W.B., Barr, A.C., 2006. Heat flux from impact ring graben on Europa. *Lunar Planet. Sci.* 37, abstract 2399.

Lipps, J.H., Rieboldt, S., 2005. Habitats and and taphonomy of Europa. *Icarus.* 177, 515-527.

Lucchita, B.K., Soderblom, L.A., 1982. The geology of Europa, in: Morrison, D. (Ed.), Satellites of Jupiter, Univ.of Arizona Press, *Tucson,* pp. 521-555.

Malin, M.C., Pieri, D.C., 1986. Europa, in: Burns, J.A., Matthews, M.S. (Eds.), Satellites, Univ. Arizona Press, T*ucson,* pp. 689-717.

Madigan, M.T, Martinko, J.M, Parker, J., 1997. Brock Biology of Microorganisms. Prentice Hall.

Marion, G.M., 2001. Carbonate mineral solubility at low temperatures in the Na-K-Mg-Ca-H-Cl-SO4-OH-HCO3-CO3-CO2-H2O system. *Geochim. Cosmochim. Acta.* 65, 1883-1896.

Marion, G.M., 2002. A molal-based model for strong acid chemistry at low temperatures (< 200 to 298 K). *Geochim. Cosmochim. Acta.* 66, 2499-2516.

Marion, G.M., Fritsen, C.H., Eicken, H., Payne, M.C., 2003. The Search for Life on Europa: Limiting Environmental Factors, Potential Habitats, and Earth Analogues. *Astrobiology.* 3, 785-811.

McCord, T.B., Hansen, G.B., Fanale, F.P., Carlson, R.W., Matson, D.L., Johnson, T.V., Smythe, W.D., Crowley, J.K., Martin, P.D., Ocampo, A.R., Hibbitts, C.A., Granahan, J.C., 1998. Salts on Europa's surface detected by Galileo's Near-Infrared Mapping Spectrometer. *Science.* 280, 1242-1245.

McKinnon, W.B., 1999. Convective instability in Europa's floating ice shell. *Geophys. Res. Lett.* 26, 951-954.

Moore, W.B., 2006. Thermal equilibrium in europa's shell. *Icarus.* 180, 141-146.

Nimmo, F., Manga, N., 2002. Causes, characteristics and consequences of convective diapirism on Europa. *Geophys. Res. Lett.* 29, 2109, 10.1029/2002GL015754.

Nimmo, F., Giese, B., Pappalardo, R.T., 2003. Estimates of Europa's ice shell thickness from elastically-supported topography. *Geophys. Res. Lett.* 30, 1233, 10.1029/2002GL016660.

Pappalardo, R.T., et al., 1998. Geological evidence for solid-state convection in Europa's ice shell. *Nature.* 391, 365-368.

Pappalardo, R.T., Head, J.W., the Galileo Imaging Team, 1999a. Europa: role of the ductile layer. *Lunar Planet. Sci.* XXX, 1967 abstract, [CD-ROM].

Pappalardo, R.T., et al., 1999b. Does Europa have a subsurface ocean? Evaluation of the geological evidence. *J. Geophys. Res.* 104, 24015-24055.

Pappalardo, R.T., Head, J.W., 2001. The thick-shell model of Europa's geology: Implications for crustal processes. *Proc. Lunar Sci. Conf.* 32. Abstract 1866.

Potts, M., 1994. Desiccation Tolerance of Prokaryotes. *Microbiol. Rev.* **58,** 755-805.

Price, P.B., 2003. Life in solid ice on Earth and other planetary bodies. In: Norris, R. and Stootman, F. (Eds.), *Bioastronomy.* 2002: Life among the stars, IAU Symposium Series No. 213, pp. 363-366.

Price, P.B., Sowers, T., 2004. Temperature dependence of metabolic rates for microbial growth, maintenance, and survival. *Proc. Natl. Acad. Sci. USA.* 101, 4631-4636.

Prockter, L.M., Schenk, P., 2005. Origin and evolution of Castalia Macula, an anomalous young depression on Europa. *Icarus.* 177, 305-326.

Prockter, L.M., Antman, A.M., Pappalardo, R.T., Head, J.W., Collins, G.C., 1999. Europa: Stratigraphy and geological history of the anti-jovian region from Galileo E14 solid-state imaging data. *J. Geophys. Res.* 104, 16531-16540.

Reynolds, R.T., Squyres, S.W., Colburn, D.S. McKay, C.P., 1983. On the habitability of Europa. *Icarus.* 56, 246–254.

Rothschild, L.J., Mancinelli, R.L., 2001. Life in extreme environments. *Nature.* 409, 1092-1101.

Ruiz, J., 2005. The heat flow of Europa. Icarus 177, 438–446.

Ruiz, J., Tejero, R., 2000. Heat flows through the ice lithosphere of Europa. *J. Geophys. Res.* 105, 23283–23289.

Ruiz, J., Tejero, R., 2003. Heat flow, lenticulae spacing, and possibility of convection in the ice shell of Europa. *Icarus.* 162, 362–373.

Ruiz, J., Montoya, L., López, V., Amils, R., 2007. Thermal diapirism and the habitability of the icy shell of Europa. *Orig. Life Evol. Biosph.* 37, 287-295.

Scheleper, C., Pühler, G., Kühlmorgen, B., Zilling, W., 1995. Life at extremely low pH. *Nature.* 375, 741-742.

Schenk, P.M., 2002. Thickness constraints on the icy shells of Galilean satellites from a comparison of crater shapes. N*ature.* 417, 419-421.

Schilling, N., Khurana, K.K., Kivelson, M.G., 2004. Limits on an intrinsic dipole moment in Europa. *J. Geophys. Res.* 109, E05006, 10.1029/2003JE002166.

Schulze-Makuch, D., Irwin, L.N., 2002. Energy cycling and hypothetical organisms in Europa's ocean. *Astrobiology.* 2, 105-121.

Spaun, N.A., Head, J.W., Pappalardo, R.T., 2004. Europan chaos and lenticulae: a synthesis os size, spacing, and areal density analyses. *Lunar Planet. Sci. Conf.* 35, abstract 1409.

Spencer, J.R., Grundy, W.M., Dumas, C., Carlson, R.W., McCord, T.B., Hansen, G.B., Terrile, R.J., 2006. The nature of Europa's dark non-ice surface material: Spatially-resolved high spectral resolution spectroscopy from the Keck telescope. *Icarus.* 182, 202-210.

Spohn, T., Schubert, G., 2003. Oceans in the icy Galilean satellites? *Icarus*. 161, 456–467.

Thomas, D.N., Dieckmann, G.S., 2002. Antarctic Sea Ice-A Habitat for Extremophiles. *Science*. 295, 641-644.

Welsh, D.T., 2000. Ecological significance of compatible solute accumulation by micro-organisms: from single cells to global climate. *FEMS Microbiol. Rev*. 24, 263-290.

Yancey, P.H., 2005. Organic osmolytes as compatible, metabolic and counteracting cytoprotectants in high osmolarity and other stresses. *J. Exp. Biol.* 208, 2819-2830.

Zahnle, K., Schenk, P., Levison, H., Dones, L., 2003. Cratering rates in the outer Solar System. *Icarus*. 163, 263-289.

Zhilina, T.N., Zavarzin, G.A., 1994. Alkaliphilic anaerobic community at pH 10. *Curr. Microbiol.* 29, 109–112.

Zolotov, M.Y., Shock, E.L., 2004. A model for low-temperature biogeochemistry of sulfur, carbon, and iron on Europa. *J. Geophys. Res.* 109, E06003, 10.1029/2003JE002194.

INDEX

A

absorption, 151, 154, 156, 169
absorption spectra, 154
absorption spectroscopy, 156
acceleration, 16, 108
accidental, 118, 120, 121, 123, 125
accountability, 5
accuracy, 89, 128
acetate, 37, 38
acetic acid, 36
acetophenone, 43, 44
acetylene, 42, 45, 47, 48
achievement, 18
acid, 29, 30, 34, 35, 36, 37, 40, 41, 43, 45, 46, 48, 50, 62, 63, 64, 65, 66, 67, 68, 69, 70, 71, 72, 73, 75, 78, 79, 89, 91, 127, 130, 134, 163, 169, 185, 186, 195, 201, 202
acidic, 35, 139
active transport, 68, 89, 119
adaptation, xi, 78, 81, 82, 88, 89, 91, 93, 94, 164, 178, 184
adenine, 35
adenosine, 71, 75
ADH, 92
adhesion, 90
ADP, 71, 75
adults, 5, 8
aerosol, x, 28, 30, 31, 32, 33, 34, 36, 39, 41, 50, 51, 53, 62
aerosols, x, 27, 28, 30, 31, 32, 33, 34, 35, 36, 37, 39, 40, 41, 44, 45, 50, 51, 77
age, xiii, 13, 19, 117, 179, 193, 194
agents, 49, 69, 71
aggregates, 51, 61, 155
aggregation, 51, 90
aggregation process, 52
air, 10, 16, 30, 31, 40, 49, 51, 52, 163
alanine, 29
alcohol, 64, 65, 67, 92
alcohols, 40, 44, 45, 77, 78
aldehydes, 40
aliens, x, 1, 12, 21, 22
alkali, 68
alkaline, 130
ALL, 60
alpha, 76, 156
alternative, 2, 10, 12, 15, 16, 17, 21, 22, 24, 28, 40, 45, 75
ambivalent, ix, 1
amide, 148
amine, 44, 63
amino, x, 27, 28, 29, 32, 35, 36, 39, 46, 47, 48, 49, 50, 51, 65, 68, 69, 76, 84, 116, 117, 128, 130, 134, 135, 139
amino acid, x, 27, 28, 29, 32, 35, 36, 39, 46, 47, 48, 49, 50, 51, 68, 69, 76, 84, 116, 117, 128, 130, 134, 135, 139
amino acids, x, 27, 28, 29, 32, 35, 36, 39, 46, 47, 48, 49, 50, 51, 68, 69, 76, 84, 116, 117, 128, 134, 135, 139
ammonia, 19, 83
ammonium, 35
amorphous, 151
amorphous carbon, 151
amphiphiles, x, 59, 61, 62, 63, 64, 65, 66, 68, 69, 71, 73, 74, 77, 79
amphiphilic compounds, 42, 129
amphiphilic systems, 62
amplitude, 125, 195
anaerobic, 40, 204
analytical techniques, 149, 188
anatase, 148
anatomy, 3
animals, 18, 19
anomalous, 203
anoxic, 157

Antarctic, xi, 81, 85, 86, 94, 95, 111, 145, 146, 149, 150, 152, 155, 165, 166, 169, 171, 172, 174, 176, 178, 190, 197, 204
anthracene, 43
aquatic systems, 199
aqueous solution, 31, 52, 65, 107, 154
aqueous solutions, 31, 107, 154
aquifers, 164
aragonite, 158, 159, 167
archaea, 82, 86, 90, 115, 117, 128
Arctic, xi, 81, 84, 85, 149, 166, 175, 176
arginine, 68, 76
argon, 46
argument, 2, 8, 10, 20, 162
Aristotle, 16
aromatic compounds, 42, 44, 45, 47
aromatic hydrocarbons, 42, 116, 117
aromatic rings, 43
aromatics, 43, 48
assessment, xii, 143, 187, 197
assignment, 164, 171
assimilation, 119
assumptions, 10, 13
astronomy, 17, 24, 25, 82, 133
astrophysics, 54, 133
asymmetric molecules, 28
asymmetry, 77, 123
Atacama Desert, 201
atmosphere, x, 5, 16, 27, 28, 29, 30, 31, 32, 39, 40, 43, 44, 45, 46, 47, 48, 54, 62, 83, 84, 133, 135
atoms, 10, 29, 120, 134, 155
ATP, 120
attachment, 38
attribution, 161, 164
Aurora, 190
authority, 6, 8, 19
availability, 48, 64, 70
awareness, 2, 8

B

bacteria, xi, 60, 81, 85, 86, 88, 91, 92, 170, 175
bacterial, xi, 81, 85, 86, 90, 91, 167, 175
bacterial cells, 86, 90
barrier, 60, 69, 71
barriers, 24, 78, 88, 89, 93
base pair, 75
behavior, 49, 74, 75, 112, 116, 140
beliefs, 24
bending, 156
benzene, 43, 44
beta-carotene, 152, 161, 171, 172, 179
bicarbonate, 39
bifurcation, xii, 51, 115, 117, 119, 120, 121, 123, 127, 130
bifurcation point, xii, 115, 118, 119, 120, 121, 123
bi-layer, 179
bioanalytical, 191
biochemistry, 3, 29
biodiversity, ix, xi, 1, 82, 86
biological activity, 84, 112
biological systems, 124, 129
biomarker, 187
biomarkers, xii, 143, 179, 184, 185
biomaterials, 150
biomolecular, xii, 143, 145, 154, 163, 190
biomolecule, 183
biomolecules, 28, 29, 62, 148, 150, 154, 160, 161, 164, 178, 182
biopolymers, 31, 56, 77, 128
biosphere, 84, 115, 128, 129, 197, 199
biosynthesis, 130
biota, 17
biotic, 116, 134
birds, 18
birth, 7, 18
blocks, x, xi, 8, 27, 28, 42, 59, 60, 68, 69, 71, 73
blood, 3
blot, 8
Bohr, 2
boiling, 52, 66, 85, 117, 197
bonding, 65, 156
bonds, 37, 44, 45, 67, 75, 155, 160, 179
bottlenecks, 9
bottom-up, 29, 61
brain, 9
branching, xi, 97, 100, 101, 110
budding, 74
building blocks, x, xi, 27, 28, 42, 59, 60, 68, 69, 71, 73

C

Ca^{2+}, 159
calcium, 156, 159, 166, 172
calcium carbonate, 159, 166
calcium oxalate, 156, 172
carbon, 29, 30, 31, 33, 35, 39, 43, 46, 64, 65, 70, 76, 83, 99, 104, 130, 151, 156, 199, 204
carbon dioxide, 31, 99, 104
carbonates, 148, 167, 178
carboxylates, 37
carboxylic, 28, 29, 30, 36, 37, 40, 41, 45, 53, 63
carboxylic acids, 28, 29, 30, 36, 37, 40, 41, 45, 53, 63
carboxylic groups, 63

carotene, 151, 152, 161, 169, 171, 172, 179, 180, 185
carotenoids, 148, 161, 175, 180
case study, 23, 24
catalase, 71
catalysis, 60, 68, 75, 76, 79, 87, 89
catalyst, 72, 73
catalytic activity, 49, 116
catalytic effect, 44
catalytic properties, 39
cation, 68
C-C, 148, 160
celestial bodies, 7, 135
cell, 60, 62, 63, 68, 71, 72, 73, 74, 77, 79, 88, 90, 119, 120, 121, 123, 124, 129, 199
cell adhesion, 90
cell division, 73, 88
cell growth, 79
cell membranes, 77
cellulose, 37
centrifugal forces, 16
cerebral hemisphere, 123
CH4, x, 27, 29, 32, 33, 36, 39, 43, 46, 47, 48, 84
changing environment, 90
channels, 83, 86
chaos, 130, 203
chaperones, 89, 90, 91, 94
charge density, 155
chemical bonds, 155, 159, 197
chemical reactions, 10, 21, 29, 61, 64, 88, 117, 125, 127, 136
chemical reactor, 77
chemical structures, 161
chemicals, xii, 61, 143
children, 8, 9, 10, 11, 13, 18, 19, 20, 21, 25
chimneys, 136
chiral, x, 27, 28, 49, 50, 51, 53
chiral catalyst, 49
chiral recognition, 49, 51
chirality, x, 27, 49, 51
chloride, 39
chlorophyll, 171, 172, 173, 174
chondrites, 139
chromatograms, 37
chromatography, 31, 43
chromosome, 123
chromosomes, 123
circulation, 119, 128, 197
classroom, 11, 12, 25
classroom practice, 25
climate change, 3, 4
clone, 86, 87
clouds, 139
C-N, 148
Co, 23, 63, 79
CO2, x, 27, 29, 33, 84, 99, 100, 104, 108, 112, 135, 136, 139, 202
coal, 43
cognitive abilities, 9
cognitive psychology, 23
coherence, 13, 16
collaboration, 190
colonisation, 148, 150, 161, 164, 166, 167, 168, 172, 174, 178, 179, 180
colonization, 173
communication, 2, 5, 8, 20, 22, 194, 196
communities, xi, xii, 81, 85, 115, 127, 128, 165, 172, 178
community, 6, 7, 86, 87, 127, 139, 166, 172, 174, 204
compatibility, 154
competition, 4, 49, 79, 120
complex systems, 134
complexity, 17, 29, 67, 69, 77, 125, 128, 129
components, x, xi, 21, 27, 37, 38, 42, 51, 59, 60, 61, 64, 67, 71, 72, 73, 74, 87, 90, 117, 121, 123, 135, 136, 151, 155, 161, 173, 179, 187
composition, 29, 31, 32, 65, 69, 74, 77, 83, 87, 88, 116, 155, 163, 166, 194, 195, 202
compounds, x, 27, 29, 30, 32, 33, 35, 36, 37, 40, 43, 44, 45, 49, 59, 62, 64, 65, 66, 76, 77, 83, 84, 117, 128, 163, 187, 199
comprehension, 17
compressive strength, 109
computation, 140
computer simulation, 69
concentrates, 124
concentration, 30, 31, 39, 40, 41, 46, 53, 62, 64, 65, 71, 75, 84, 119, 125, 154, 155, 197
conception, 10, 15, 16, 17, 116, 121
condensation, x, 27, 30, 31, 32, 77
conductive, xiii, 193, 194, 195, 196
conductivity, 107, 119
confidence, 2
configuration, x, 27, 51, 105
conjugated dienes, 44
conjugation, 88, 160
consciousness, 7
consensus, 3, 6
conservation, 123
conspiracy, 21
constraints, 88, 145, 203
construction, xi, 59, 74, 154, 160, 187
contaminants, 49, 51
contamination, 46, 50
control, 32, 33, 34, 35, 36, 37, 39, 40, 44, 46, 74, 129
control condition, 34

convection, 196, 203
convective, 195, 196, 202
conversion, xii, 71, 72, 73, 115, 125, 128
cooling, 61, 66, 70, 107, 136, 139
copper, 3
correlation, 40, 109, 120, 121, 153
cosmic rays, 29
coupling, 79
covalent, 67, 75
covalent bond, 75
covering, 108, 200
crack, 166, 167
cracking, 111
credentials, 6
critical state, 121
critical thinking, 23
cross-sectional, 144
crust, 112, 166, 167, 172, 173, 180, 196, 198, 202
crystal growth, 52
crystal structure, 39, 92, 156
crystal structures, 156
crystalline, 49
crystallites, 50
crystallization, 49, 50, 51, 52
crystals, 3, 7, 30, 49, 50, 51, 52, 90, 176, 180, 197
culture, ix, 1, 12, 21, 22, 23, 86
culture conditions, 86
curriculum, 9, 12, 24
customers, 7
cyanide, 35, 40, 47
cyanobacteria, 86, 165, 166, 172, 178, 179
cycles, x, 28, 31, 36, 42, 43, 44, 45, 46, 47, 48, 49, 60, 65
cycling, 31, 42, 203
cysteine, 148
cysteine residues, 148
cytoplasmic membrane, 89
cytosine, 46, 47
cytosol, 94

D

danger, 5, 17
database, 160, 163, 164, 184, 188
daughter cells, 73, 74
dead zones, 65
death, 4, 94
debates, 2, 3, 4, 23
decision making, 4
decisions, 3
decomposition, 47, 48, 76, 120, 133, 136
deep-sea hydrothermal vents, 94, 149
definition, 156, 162, 166
degradation, 46, 71, 91, 118, 125, 151, 152, 153
dehydration, 30, 53, 65, 77, 197
dehydrogenase, 68, 92
democracy, 3
denaturation, 88, 90
density, xiii, 37, 105, 138, 139, 160, 193, 194, 195, 203
deposition, 110
deposition rate, 110
deposits, 166, 176
depression, 104, 105, 107, 203
derivatives, 43, 44, 46, 47, 62, 63, 65, 67, 77, 148
desert, 145, 171, 180
desiccation, xii, 143, 149, 156, 164, 197
desorption, 91
destruction, 109, 118, 119, 121
detection, xii, 52, 89, 143, 149, 150, 155, 156, 166, 179, 183, 188
detritus, 161
deuterium oxide, 154
developed nations, 5
developing nations, 2
diagenesis, 129
dialysis, 37, 38
dichotomy, 122, 124, 125
differentiation, 159
diffusion, 60, 68, 71, 111, 119
diode laser, 144, 150, 153, 154, 188
dipole moment, 156, 203
disaster, 3
discharges, x, 3, 27, 28, 29, 30, 36, 40, 41, 42, 43, 44, 46, 47, 48, 117, 127
discrimination, 51, 159, 160, 163, 164, 167, 183
disorder, 94
dispersion, 52
disseminate, 5
dissipative structure, 117, 118, 123
dissipative structures, 117, 118, 123
distribution, x, 5, 28, 37, 73, 77, 137, 139
divergence, 123
diversity, 3, 29, 35, 37, 53, 84, 86, 155, 160, 178
division, 72, 73, 74, 76, 79, 88, 120, 121, 123
DNA, xii, 67, 68, 69, 70, 84, 86, 87, 88, 89, 91, 92, 115, 120, 127, 128, 198
dogmas, 17
do-it-yourself, 8
double bonds, 37, 88
drought, 65
duration, 135
dust, 62, 99, 100, 110
dyes, 43, 68, 69
dynamic environment, 53

E

earth, ix, 1, 4, 5, 13, 14, 16, 21, 62, 77, 133, 190
Earth Science, 97, 129
ecological, xii, 115, 127, 128, 129, 194
ecological systems, xii, 115, 127, 128
ecologists, 128
ecology, xi, 3, 82, 86, 202
ecosystem, 4, 94, 130, 199
ecosystems, xii, 60, 84, 86, 93, 115, 128, 134, 199, 201
education, ix, 1, 18, 22, 23, 24, 25, 131, 140, 141
educational research, 12, 22
electric charge, 179
electric field, 119
electrodes, 42
electromagnetic, 135, 144, 150, 159
electron, 10, 86, 179
electrostatic interactions, 63, 65, 67, 69
elongation, 92, 128
embryos, xii, 115, 128
emission, 144, 145, 146, 150, 151, 152, 153, 171, 187
enantiomer, 49
enantiomers, 28
encapsulated, 68, 70, 71, 72
encapsulation, 66, 67, 69, 77
energy, ix, xi, 1, 3, 29, 30, 44, 46, 48, 49, 53, 59, 60, 61, 64, 66, 68, 72, 74, 75, 92, 94, 119, 120, 124, 125, 126, 128, 135, 144, 150, 151, 199
energy supply, 199
entropy, 124, 125
environment, xii, 29, 30, 31, 36, 42, 50, 60, 62, 64, 70, 72, 74, 84, 90, 92, 97, 111, 116, 119, 120, 124, 133, 135, 136, 139, 140, 149, 160, 197, 198
environmental change, 84
environmental conditions, xi, 53, 60, 61, 64, 81, 85, 145
environmental effects, 159
environmental factors, xi, 60, 62, 94
enzymatic, 64, 71, 89
enzymes, xi, 2, 71, 79, 81, 88, 89, 90, 94, 127
epistemology, 23
equilibrium, 44, 52, 116, 118, 121, 123, 128, 202
equilibrium state, 123
erosion, 109, 110, 164
Escherichia coli (E. coli), 91, 94
ethylene, 160
European Space Agency, xii, 143, 188, 189
evaporation, x, 27, 30, 39, 50, 51
evolution, x, xi, xii, 1, 2, 5, 17, 18, 19, 20, 21, 23, 29, 31, 42, 47, 49, 54, 59, 67, 73, 76, 77, 82, 84, 85, 89, 90, 93, 94, 115, 117, 127, 128, 129, 130, 132, 134, 135, 139, 161, 181, 201, 203
evolutionary process, 60
examinations, 10, 18, 19
excitation, 144, 146, 148, 150, 151, 152, 153, 154, 155, 158, 161, 163, 171, 174, 181, 182, 183, 184, 187, 188
exercise, 137
ExoMars, xii, 143, 183, 188, 189, 190
exopolysaccharides, 90, 94
experimental condition, 29, 30, 40, 49, 51, 53, 145
expertise, 8, 60, 190
exploitation, ix, 1
exposure, 66, 70, 89, 90, 101
external environment, 60
external influences, 116, 125
extraction, 52, 116, 149
extraterrestrial, 84
extremophiles, xii, 82, 90, 94, 95, 143, 164, 166, 184, 188
eyes, 123

F

failure, 6, 109
fatty acids, x, xi, 59, 62, 63, 64, 65, 66, 67, 68, 81, 82
ferrous ion, 39
film, 106, 128
films, 112, 163
filtration, 52
fire, 13, 14
fission, 79, 121, 123
flexibility, 21, 89, 93
flight, 91
floating, 16, 202
flood, 91
flow, 52, 99, 100, 101, 104, 107, 108, 109, 173, 178, 196, 203
fluctuations, 49, 117, 119, 120, 127, 128
fluid, 79, 117
fluorescence, 91, 144, 145, 146, 150, 151, 152, 153, 171, 187
fluorine, 10
fluorine atoms, 10
focusing, 13, 33, 132
folding, 65, 88, 89, 94
foodstuffs, 3
formaldehyde, 29
fossil, 19, 66, 178
fractionation, 37
fracture, 166
fractures, xiii, 193, 194

fragmentation, 43
free energy, xii, 115, 116, 124, 125
freeze-dried, 33, 46
freezing, xi, 44, 45, 81, 85, 89, 104, 105, 107, 109, 197
Freud, 6
friction, 10, 107, 108
frost, 98, 100, 101, 104, 109, 110, 136, 139
funding, 20, 188
fungi, 86
fusion, 67, 73

G

galaxy, 135
Galileo, xi, 81, 83, 194, 201, 202, 203
games, 132, 141
gas, 29, 30, 31, 35, 37, 39, 40, 43, 47, 101, 198
gas chromatograph, 31, 43
gas jet, 101
gas phase, 29, 37, 40
gases, 29, 62, 84, 150, 174
gene, 89, 123
general knowledge, xii, 131, 132, 133
generation, 29, 44, 46, 47, 48, 64, 71, 72, 75, 144, 145, 148, 150
genes, 18, 72, 86, 88, 89, 91, 127
genetic code, 68, 130
genetic information, 70
genome, 89, 123, 128
genome sequences, 89
genomic, 90, 91
genomics, 88, 91
genotype, 94, 127
geochemistry, 129, 190
geological history, 203
geology, 83, 154, 159, 190, 194, 200, 202, 203
geothermal, 85, 117, 136
geyser, 99, 100
glaciers, 84, 85, 86, 87
global warming, 3
glucose, 68
glutamine, 92
glycerol, 64, 66, 67, 69, 71, 75, 77, 197
glycine, 29, 46, 49, 117, 197
grain, 105, 108, 110
grains, xi, 84, 97, 99, 104, 107, 108, 110, 155, 157
gravitational field, 16
gravity, 10, 16, 101
greenhouse, 4, 84
greenhouse gases, 84
GroEL, 89, 90, 92, 93
groups, 33, 36, 37, 44, 63, 92, 132, 139, 149, 154, 160, 183
growth, x, xi, 52, 59, 71, 72, 73, 74, 76, 78, 79, 81, 85, 88, 91, 203
growth temperature, 91

H

H_2, 10, 29, 32, 43, 46, 47
habitat, 88, 95, 144, 146, 199
halophiles, 198
Hamaker constant, 105, 106
handedness, 50
hands, 132
hanging, 3
harvest, 72
harvesting, xi, 59, 72, 74, 75, 170, 174
hazards, 188
healing, 7
heat, xi, 29, 82, 89, 90, 91, 93, 94, 105, 107, 119, 136, 196, 203
heat conductivity, 119
heat shock protein, xi, 82, 89, 91, 93, 94
heating, xiii, 46, 109, 150, 193, 195, 196
heavy metal, 164
heavy water, 154
height, 139
Heisenberg, 2
helium, 154
hemisphere, 98, 100, 196, 202
Hendra, 161, 191
heredity, 123
heterocycles, 36, 37, 46
heterogeneity, 119, 160
heterogeneous, 119, 153, 155, 161
heteropolymers, 37
Higgs, 113
high pressure, 31, 44, 62, 73, 197
high school, 10
high temperature, 88, 90, 91, 93, 117, 128, 198
higher education, 132
high-frequency, 42
holistic, 21
homochirality, x, 27, 28, 30, 49, 51
homopolymers, 75
host, xii, 71, 143, 148
hot spring, 52, 85, 117, 127
HPLC, 37, 38
HSP90, 90, 91, 92, 93
Hsps, xi, 82, 89, 90, 91, 93
Hubble, 5
human, 3, 7, 11, 17, 19, 21, 166, 188
human actions, 3

humans, 3, 4, 18, 164
humidity, 107
hyaline, 173
hydrates, 111, 156, 195
hydration, 53
hydro, xi, 28, 33, 35, 36, 37, 38, 40, 42, 44, 51, 53, 59, 62, 66, 76, 83, 104, 116, 117, 148, 150
hydrocarbon, 37, 62, 63, 64, 66, 68, 69, 70, 72, 75, 79, 148, 156
hydrocarbons, 38, 40, 42, 62, 66, 76, 83, 148
hydrogen, 10, 65, 137, 156
hydrogen bonds, 65
hydrological, 164
hydrolysis, 35, 39, 83
hydrophilic, 33, 34, 35, 36, 37, 38, 44, 51, 104
hydrophilic groups, 37
hydrophobic, 30, 33, 34, 38, 44, 50, 51, 65, 69, 72, 75
hydrophobic anchors, 72
hydrosphere, 54, 128
hydrostatic pressure, 85
hydrothermal, xiii, 61, 76, 84, 117, 127, 128, 129, 130, 194, 197, 199
hydrothermal activity, 199
hydrothermal system, 117, 127
hydroxyapatite, 148
hydroxyl, 30, 43, 48, 65, 69, 149, 154
hydroxyl groups, 149, 154
hypothesis, x, 27, 32, 36, 37, 40, 42, 45, 60, 65, 69, 71, 84, 89, 113, 195
hysteresis, 90

I

ice, xii, xiii, 30, 31, 42, 44, 45, 46, 47, 48, 53, 83, 84, 86, 87, 88, 90, 94, 97, 99, 100, 104, 105, 107, 108, 109, 110, 111, 112, 135, 136, 149, 154, 156, 178, 193, 194, 195, 196, 197, 198, 200, 202, 203
identification, xii, 143, 155, 161, 163, 175, 178, 184, 187
identity, 8
illumination, 98, 100, 144, 155
imagery, 130
images, xi, 15, 16, 23, 97, 98, 99, 100, 102, 103, 164, 201
imagination, x, 1, 12, 17
imaging, 176, 202, 203
imbalances, 49
impact craters, xiii, 136, 166, 193, 194
imprinting, 11
imprisonment, 6
impurities, 50, 51, 198
in vitro, 125
inclusion, xii, 143, 175, 176, 183
indicators, xii, 135, 143, 153, 188
individuality, 67, 68
induction, 49, 51, 88, 90
industrial, 3, 60
information exchange, 125
information processing, 9, 88
infrared, 107, 144, 145, 150, 151, 153, 154, 156, 161, 171, 183, 201
infrared spectroscopy, 156
inhospitable, 82, 83, 166, 173
initial state, 118, 120, 123
inorganic, x, xii, 27, 30, 39, 49, 62, 78, 143, 144, 148, 149, 178, 187, 199
inorganic salts, 39
insight, 132, 133
instabilities, 68
instability, 64, 74, 117, 121, 123, 202
instruments, 8, 153, 163, 183
integration, 24, 61, 131, 162
integrity, 119, 123, 154
intelligence, 12, 135
interaction, xii, 12, 42, 49, 62, 64, 65, 66, 67, 68, 69, 73, 75, 99, 105, 115, 116, 120, 127, 149
interactions, xi, 60, 63, 64, 65, 66, 67, 68, 69, 73, 74, 75, 76, 82, 88, 91, 93, 159
interdisciplinary, 22, 82, 132, 134
interface, x, 28, 30, 31, 35, 37, 40, 42, 49, 51, 52, 65, 107, 163, 172, 180, 199
interfacial layer, 108
interplanetary, 62
interview, 10, 13, 23
intrinsic, 84, 144, 203
intuition, 10
inversion, 125, 130
inversions, 121
investment, ix, 1
ionic, 61, 68, 78, 119
ionization, 46, 91
ionizing radiation, 30, 198
ions, 30, 65, 68, 75, 78, 120, 144, 148, 156, 159
IR spectroscopy, 36
iron, 39, 76, 199, 204
irradiation, 30, 33, 36, 39, 61, 151, 154, 155
ischemia, 94
ischemic, 94
ischemic tolerance, 94
island, 18, 174
isolation, 46
isomers, 28
isotope, 129
isotopes, 130
isotopic, 129

K

kerogen, 44, 45
Keynes, 24
Krebs cycle, 36

L

L1, 92
L2, 42
lakes, xi, 31, 81, 83, 84, 85, 149, 164, 166, 178, 179
lambda, 161, 162
land, 65
language, 23
laser, 91, 144, 146, 150, 151, 152, 153, 154, 155, 161, 162, 163, 171, 174, 175, 176, 181, 182, 183, 184, 187, 188
laser radiation, 144, 151, 162, 163, 183
laws of motion, 16
leakage, 69, 70, 73
learners, 2, 8, 9, 11, 12, 13, 16, 20, 25
learning, x, 1, 2, 5, 9, 11, 12, 13, 17, 19, 20, 21, 22, 23, 24, 25
learning difficulties, x, 1, 12, 20, 22
lens, 161, 162, 163, 181
lenses, 161
LIBS, xii, 143, 188, 189
lichen, 144, 146, 163, 184
life cycle, 62, 72
life forms, 19, 60
lifetime, xi, 59
ligands, 89
likelihood, 73, 87
limitations, xi, 60, 66, 70, 184
linear, xiii, 16, 17, 20, 37, 109, 127, 154, 193, 194, 200
linear dependence, 154
linkage, 72, 89, 156
links, 20, 120
lipid, 60, 66, 76, 78, 88, 148
lipids, 28, 60, 65, 66, 74, 78, 88, 116
lipophilic, 72
liposome, 68
liposomes, 79, 116, 117
liquid interfaces, 39
liquid phase, 31, 44, 48, 50
liquid water, x, xii, xiii, 27, 30, 31, 32, 33, 34, 35, 44, 61, 82, 83, 97, 98, 104, 105, 107, 109, 110, 111, 112, 133, 135, 136, 137, 139, 193, 194, 198
lithosphere, 196, 201, 203
local order, 53
localised, 156
location, 65, 98, 99, 110, 139, 144, 146, 153, 155, 162
long period, 87
losses, 5, 43
low molecular weight, 40
low power, 32
low temperatures, 42, 47, 63, 84, 88, 89, 90, 93, 94, 166, 202
low-density, 195
low-temperature, 204
LPS, 140
luminosity, 84
lying, 150, 195

M

macromolecules, 68, 85, 87, 116
magnesium, 159, 166, 178, 195
magnetic field, xiii, 193, 194
mainstream, 20
maintenance, x, 59, 70, 71, 87, 121, 123, 127, 203
malate dehydrogenase, 92
malic, 36, 40
manifold, 11, 24, 120, 128
manipulation, 187
mapping, 54, 153, 202
marine environment, 94
market, 3
Mars, xi, xii, 30, 47, 54, 82, 83, 97, 98, 99, 104, 107, 110, 111, 112, 113, 134, 135, 136, 137, 138, 139, 140, 143, 145, 149, 151, 164, 165, 166, 171, 172, 174, 178, 179, 183, 184, 188, 189, 190, 191
Martian, xii, 101, 104, 109, 110, 111, 112, 136, 143, 155, 160, 164, 174, 179, 184, 188
Marx, 94
Marxism, 6
Marxist, 6
mask, 187
mass media, x, 1, 141
mass spectrometry, 43, 91
maternal, 121
matrix, xii, 47, 91, 109, 133, 134, 135, 136, 139, 143, 148, 149, 151, 155, 156, 159, 174, 175, 178, 180
maturation, 91
MDH, 92
measurement, 162
measures, 37
media, x, 1, 7, 12, 67, 68, 74, 130, 141
mediators, 42
meiosis, 122, 124
melamine, 46

melt, 31, 42, 43, 44, 45, 46, 47, 48, 49, 83, 84, 86, 176, 178, 195
melting, x, 28, 44, 64, 76, 104, 105, 107, 108, 111
melting temperature, 105, 107, 108
meltwater, 87
membranes, xi, 37, 38, 62, 66, 68, 69, 72, 73, 75, 77, 78, 79, 81, 88, 119, 120
memory capacity, 9
mental model, 10
mercury, 156
mesophilic organisms, 89
metabolic, 60, 62, 64, 67, 71, 72, 73, 74, 84, 89, 127, 197, 198, 203, 204
metabolic pathways, 89
metabolic rate, 203
metabolic systems, 74
metabolism, 36, 60, 65, 68, 70, 72, 84, 87, 92, 116, 125, 197, 199
metal content, 135
metal ions, 75
metals, 30, 56, 68
metaphor, 4, 11, 23
meteor, 74, 83
meteorites, 62, 63, 77, 134, 135
methane, 31, 40, 45, 46, 47
Mg^{2+}, 65, 71, 159
micelles, 73
microalgae, 86
microbes, 85, 166, 178
microbial, xi, xii, 46, 76, 81, 82, 83, 84, 85, 86, 87, 93, 94, 115, 128, 130, 165, 166, 169, 197, 201, 203
microbial activity, 86
microbial communities, 165
microbial community, 87
microcosm, 11
microenvironment, 62
microenvironments, 47, 110
micro-environments, 30
microorganism, 93
microorganisms, xi,12, 81, 82, 84, 85, 86, 88, 90, 93, 94, 129, 171, 198, 201, 202, 204
microscope, 153, 160, 161, 162
microscopy, 155, 161, 162, 163
microspheres, 116
microsystem, 116, 121, 124, 125, 126
mineralogy, 159
minerals, 30, 159, 166, 173, 188, 195
minority, 2, 5
mirror, x, 27, 49
misconception, 17
missions, xi, 4, 81, 83, 93, 154, 188, 190
mitosis, 122, 124
mobility, 53
model system, xi, 59
modeling, 67
models, xi, 9, 10, 12, 32, 49, 72, 74, 76, 79, 81, 84, 85, 93, 104, 116, 128, 164, 179, 188, 195, 199
modules, 131, 132
modulus, 109
moieties, 69, 148, 156
molecular biology, 82
molecular dynamics, 160
molecular mechanisms, xi, 81, 82, 88, 93
molecular structure, 144, 156
molecular weight, 37, 40, 69
molecular weight distribution, 37
monolayers, 106, 108
monomers, 45, 73
montmorillonite, 108
Moon, 5, 194
morphological, 98, 111, 121
morphology, 86, 98, 101
motion, 10, 16, 17, 20, 24, 101, 121
motivation, 2, 11, 21
mountains, xi, 81, 83, 85, 86
movement, 6, 8, 23, 99, 101, 103, 104, 120, 197
mRNA, 89
multicellular organisms, 123
multidisciplinary, 132

N

Na^{+}, 159
NaCl, 39
NAD, 92
nano-crystals, 50
naphthalene, 43, 44, 117
NASA, 5, 83, 94, 113, 140, 181, 182, 183, 189
National Research Council, 132, 141
natural, ix, x, 1, 3, 7, 17, 18, 19, 20, 21, 28, 31, 39, 49, 50, 57, 69, 74, 88, 117, 118, 125, 127, 128, 129, 131, 132, 136, 139, 166
natural habitats, 88
natural resources, ix, 1
natural sciences, 131, 132, 139
natural selection, 17, 18, 19, 20, 21, 57, 74, 127
neglect, 60
neon, 154
network, 71, 72, 101, 119, 125, 127
Newton, 16, 192
Newtonian, 10, 16
nitrate, 148
nitrogen, 31, 33, 40, 45, 46, 47
NMR, 36, 37
non-biological, 62, 63

non-destructive, xii, 143
non-enzymatic, 31, 70, 71
nonequilibrium, 116, 117, 118, 119, 120, 121, 130
nonequilibrium systems, 130
normal, 67, 123, 127, 151, 171, 197
nuclear power, 3
nucleation, x, 28, 51
nucleic acid, x, 27, 28, 29, 45, 69, 70, 71, 72, 73, 75, 84, 91, 116, 127
nucleoprotein, xii, 115, 127
nucleotides, 68, 69, 70, 71, 84, 117
nucleus, 119, 120
numerical aperture, 161, 162, 163
nutrient, 72, 86
nutrients, 60, 71, 73, 88, 90, 164
nutrition, 68

O

observations, 30, 40, 73, 100, 139, 194, 202
oceans, x, xii, 27, 30, 31, 54, 61, 64, 65, 66, 84, 85, 131, 138, 139, 194
oil, 112, 179
oleic acid, 63, 66, 68, 69, 70, 71, 73
oligomers, 37, 69, 70, 73
optical, 109, 144, 154, 162, 181, 182
optical properties, 109
orbit, 17
organic compounds, x, 27, 28, 29, 36, 39, 44, 53, 55, 62, 117
organic matter, 31, 33, 62, 63, 116, 128
organism, x, 27, 66, 87, 90, 115, 119, 120, 121, 123, 124, 125, 163, 171, 172, 197, 198
origin of chirality, 28, 54
oscillation, 121
oscillations, xii, 115, 121, 123, 128
osmotic, 73
oxalate, 76, 156, 172
oxidants, 110, 169
oxidation, 40, 44, 63, 66, 199
oxide, 159
oxygen, 5, 117, 129, 137
ozone, 172

P

PAHs, xi, 28, 38, 42, 43, 44, 45, 46, 59, 62, 72
palliative, 71
paradoxical, 120, 123
parameter, 61, 156, 163, 187
particles, 31, 32, 43, 44, 62, 120, 155
passive, 101, 102
pathways, 36, 42, 48, 73, 89
pedagogy, 12
penumbra, 98
peptide, 28, 37, 78
perceptions, 4, 21, 22
perchlorate, 111
performers, 8
periodic, 123, 125
permafrost, xi, 81, 82, 84, 85, 86, 107, 149
permeability, 66, 68, 69, 70, 78
permeation, 68, 69
petroleum, 60
pH, 33, 34, 39, 41, 53, 63, 64, 65, 66, 68, 78, 85, 132, 135, 164, 197, 198, 203, 204
pH values, 85
phase diagram, 136, 139
phenotype, 94, 127
phenotypic, 88
philosophers, 7
philosophy, 17, 21
phosphate, 66, 68, 78
phosphates, 65, 77
phospholipids, 62, 63, 64
hotocatalytic, 79
photochemical, 42, 151
photochemical degradation, 151
photosynthesis, 110, 178
photosynthetic, 172, 179, 199
phycocyanin, 174, 185
phylogenetic, 84, 86, 87
phylogenetic tree, 87
phylum, 86
physical properties, 88
physics, 9, 10, 16, 23, 133, 139
physiological, 90
physiology, 3, 87, 89, 93
pigments, 169, 179, 180
pinhole, 162
plagioclase, 160
planetary, 4, 29, 30, 44, 45, 47, 82, 132, 133, 134, 135, 136, 137, 138, 139, 140, 145, 150, 154, 155, 161, 164, 166, 183, 184, 187, 188, 203
planets, 13, 14, 16, 17, 29, 45, 82, 83, 85, 129, 133, 134, 135, 139
plasma, xi, 30, 81
plasma membrane, xi, 81
Poisson, 109
polar groups, 37
polar ice, 112
polycyclic aromatic compounds, 42
polycyclic aromatic hydrocarbon, xi, 28, 53, 59, 62, 75, 150
polymer, 37, 43, 44, 60, 69, 72, 78

polymer chains, 44
polymerase, 92
polymerization, 35, 45, 47, 48, 62, 64, 70, 71, 78, 128
polymerization mechanism, 45
polymerization process, 78
polymerization processes, 78
polymers, x, 27, 37, 42, 44, 45, 47, 59, 68, 69, 71, 72
polynucleotide, 75, 128
polyphosphates, 77
polyunsaturated fat, xi, 81, 82
polyunsaturated fatty acid, xi, 81, 82
polyunsaturated fatty acids, xi, 81, 82
population, 4, 5, 6, 7, 70, 73, 115, 150
pore, 138, 166
pores, 68, 73, 109, 174
porous, 73, 107
porphyrins, 150, 178, 179
positive interactions, 74
post-compulsory education, 5
potassium, 78
power, 3, 32, 70, 144, 150, 153, 154, 161
precipitation, 39, 40, 66
pressure, 29, 30, 31, 44, 62, 73, 82, 85, 104, 105, 109, 119, 135, 136, 137, 164, 198
prior knowledge, 15, 24
probability, 50
probe, 149, 181, 182, 183, 197
problem-solving, 12
production, 28, 29, 35, 36, 39, 42, 44, 49, 51, 62, 71, 72, 74, 77, 125, 159, 182
professional teacher, 12
program, 87, 140
programming, 125
prokaryotes, xi, xii, 81, 88, 115, 116, 127, 128
prokaryotic cell, 128
protection, 62, 140, 164, 176
protein, 2, 71, 82, 88, 89, 90, 91, 92, 93, 94, 120, 156
protein analysis, 91
protein denaturation, 90
protein folding, 88, 89
protein function, 89
protein structure, 71, 90
protein synthesis, 89, 91, 92
proteins, x, xi, 27, 28, 29, 60, 67, 68, 72, 82, 88, 89, 90, 91, 92, 93, 127, 148, 156
proteobacteria, 86, 91
proteomics, 88, 91
protocells, x, 59, 60, 79
protons, 68, 78
protoplasm, 119
prototype, 154, 181, 183, 188
Pseudomonas spp, 86
psychiatrists, 6
psychoanalysis, 6
psychological processes, 25
public, ix, 1, 2, 3, 4, 5, 7, 8, 12, 13, 20, 21, 22, 23, 132, 141
pure water, 39, 40
purines, 35, 36, 46
pyrene, 43
pyrimidine, 46
pyrite, 39, 62

Q

qualitative research, 25
quality control, 90
quality of life, ix, 1
quanta, 11, 24
quartz, 49, 62, 147, 148, 167, 168, 172, 181, 183
quinone, 179

R

radiation, x, xii, 27, 30, 82, 107, 110, 135, 143, 144, 145, 150, 151, 152, 153, 156, 158, 159, 162, 163, 164, 167, 172, 175, 176, 178, 181, 183, 191, 198
radiation detectors, 150
radical mechanism, 35
radius, 68, 108, 120
Raman scattering, 144, 149, 150, 151, 153, 154, 161, 187
Raman spectra, xii, 143, 144, 145, 146, 148, 149, 150, 151, 152, 153, 154, 155, 156, 157, 158, 159, 160, 161, 162, 163, 164, 166, 167, 169, 171, 173, 174, 175, 176, 180, 181, 182, 183, 188
Raman spectroscopy, xii, 143, 144, 146, 149, 153, 154, 155, 156, 159, 160, 161, 166, 172, 173, 179, 183, 187
random, 7, 19, 74
range, xi, xii, 5, 6, 7, 9, 11, 12, 16, 18, 20, 30, 32, 37, 51, 64, 65, 74, 77, 81, 82, 85, 88, 93, 106, 108, 117, 120, 132, 136, 144, 145, 148, 150, 153, 158, 164, 171, 182, 183, 184, 190, 195, 196, 198
reactant, 10
reactants, 44, 48, 53
reaction mechanism, 37
reaction rate, 31
reaction time, 44
reactive groups, 44
reactivity, 31, 37, 53
reagents, 31
reality, 60, 101, 133, 136, 139
recognition, 49, 51, 154, 156, 160, 174, 179, 188

refractive index, 163
refractory, 60
regolith, 30, 111, 174, 188
regular, 50, 116, 121, 123, 125, 139, 154
regulation, 89, 120
relaxation, 201
relevance, xii, 13, 45, 50, 63, 77, 94, 129, 130, 144, 146, 149, 162
remote sensing, 181, 182
renewable energy, ix, 1
reperfusion, 94
replication, 69, 70, 71, 75, 91, 116, 123
repression, 89
reprocessing, 3
reproduction, 60, 73, 75, 116
reservoirs, xiii, 193, 194, 198, 199
residues, 148, 166
resistance, 10, 16, 92, 198
resolution, xi, 97, 98, 99, 103, 110, 158, 163, 164, 183, 184, 187, 203
resources, ix, 1, 4, 23
responsibilities, ix, 1
retention, 69, 90, 128
rhythms, 121, 123
ribose, 69, 78
ribosomal, 92
ribosome, 91
ribosomes, 89
ribozyme, 65, 71
ribozymes, xi, 59, 71
rings, 44
RNA, xii, 45, 60, 65, 69, 71, 75, 76, 79, 84, 88, 89, 91, 92, 115, 116, 127, 129, 130
robotic, 145, 150, 154, 166, 183
rocky, xiii, 133, 139, 194, 195, 196, 199
room temperature, 44, 45, 46, 48, 63, 66, 67
rovers, 145, 154, 183, 188
rutile, 148

S

safeguard, ix, 1
saline, 39, 40, 41
salinity, 39, 61, 65, 76, 82, 164, 198
salt, 39, 63, 64, 66, 71, 137, 179
salts, xiii, 30, 39, 40, 65, 66, 75, 104, 107, 174, 193
sample, 43, 144, 155, 161, 162, 163, 166, 174, 181, 183, 187
sampling, xii, 143, 179
sand, 100
satellite, 194
saturated fatty acids, 63, 66
saturation, 198
scanning electron, 86
scarcity, 194
scatter, 162
scattering, 144, 149, 150, 151, 153, 154, 155, 156, 161, 187
scavenger, 47
school, 2, 3, 5, 8, 9, 10, 11, 12, 13, 16, 17, 18, 19, 20, 21
schooling, 13, 18
science education, x, 1, 2, 5, 8, 11, 13, 20, 22, 23, 140
science educators, 8
science teaching, 8, 9, 10, 11, 12, 19
scientific community, 6, 139
scientific knowledge, 3, 9
scientific theory, 20
sea floor, xiii, 194, 197, 199
sea ice, xi, 81, 85, 94, 95
sea level, 166
sea-ice, 84
search, xii, 12, 74, 82, 83, 85, 93, 94, 143, 163, 164, 188, 194, 201
searching, 188
seawater, 117, 129
secondary students, 17
sediment, 155
sediments, 149, 157, 166, 174
self-assembly, 42, 45, 61, 62, 76, 78
self-awareness, 8
self-organization, 118, 119
self-organizing, 36
self-reproduction, 116
SEM, 34, 51
sensing, 181, 182, 183
sensitivity, 3, 63, 65, 66, 120, 144, 150, 154, 155, 161, 183
sensors, 94
separation, 69, 74, 121, 149
septum, 92
sequencing, 91
series, 15, 39, 84
shape, 18, 19, 102, 133
shock, xi, 31, 82, 89, 90, 91, 93, 94, 95
shock waves, 31
short period, 70, 127
short-range, 30, 51
shoulder, 166
siderite, 39, 41
sign, 3, 7, 51, 53
signalling, 90
signals, 154, 172, 174, 188
signs, 7, 101, 116, 146, 164, 178
silica, 181, 182

silicate, 195
silicates, 148
silicon, 117, 129, 163
silver, 155
similarity, 86
simulation, 39, 129
simulations, 32
SiO2, 185
sites, xii, 108, 143, 144, 145, 154, 156, 178, 179
skeptics, 7
small heat shock proteins, 94
smokers, 127, 136
sodium, 46, 50, 51, 52, 77, 195
software, 95, 98, 140
soil, xi, 81, 85, 86, 107
soils, 106
solar, 5, 29, 62, 83, 100, 103, 110, 134, 137, 178
solar system, 5, 29, 62, 83, 137
solid phase, 43, 52
solid state, 197
solid-state, 202, 203
sols, 101
solubility, 39, 44, 202
solvent, 51, 61
space exploration, 82, 166
space station, 4
space-time, 15
spark discharges, 28, 29, 36, 40, 41, 42, 43, 44, 46, 47, 48
spatial, xi, 50, 97, 98, 99, 103, 110, 175
spatial information, 175
specialization, 132
speciation, 43
species, xii, 3, 4, 10, 18, 44, 49, 52, 62, 72, 83, 86, 91, 115, 117, 128, 150, 151, 154, 161, 183
specific surface, 107
specificity, 156, 159, 172
spectral signatures, xii, 143, 166, 171, 188
spectroscopy, 36, 42, 144, 149, 153, 155, 156, 159, 203
spectrum, 37, 144, 145, 148, 149, 150, 151, 152, 153, 154, 155, 156, 159, 160, 161, 163, 166, 167, 168, 171, 172, 173, 174, 175, 176, 177, 178, 179, 180, 182, 183, 187, 188
speed, xi, 16, 17, 73, 97, 99, 101, 110, 153, 184
Sputnik, 2
SSB, 92
stability, 42, 47, 63, 64, 65, 66, 67, 68, 69, 71, 74, 77, 88, 89, 123, 126, 127, 128, 135
stabilization, xii, 36, 69, 89, 115, 130
stabilize, 40, 65, 77
stabilizers, 67
stable states, 118, 121
stages, 9, 49, 62, 70, 128, 130
standards, 184
stars, 13, 14, 15, 133, 203
starvation, 4, 6
STEM, 2
steric, 88
sterile, 43, 117
stimulus, 8
stochastic, 51
storage, 3, 91
storms, 29
strategies, xii, 12, 88, 112, 143, 164, 166, 176, 184, 197
strength, 16, 20, 61, 109
stress, xi, 73, 90, 97, 109, 110, 125, 178
stretching, 148, 156, 159, 182
stromatolites, 177, 178
students, xii, 5, 7, 8, 9, 10, 11, 12, 13, 14, 16, 17, 19, 20, 21, 131, 132, 133, 134, 135, 136, 137, 138, 139, 140
styrene, 44
subgroups, 101
subjective, 6
substances, 29, 37, 105, 109, 197
substitution, 154, 160
substrates, 91, 150
subtraction, 182
Succinic, 35, 36, 41
sugar, 68
sugars, x, 27, 28, 29, 30, 68, 69
suicide, 6
sulfate, 111, 148, 195
sulfur, 30, 39, 199, 201, 204
sulfuric acid, 195
summer, 98, 135, 171
sunlight, 104, 166, 194, 199
supernatural, 8
superposition, 182
supramolecular, 130
supramolecular chemistry, 130
surface layer, 105
surface tension, 105
surface water, 139
surfactant, 63, 66
surfactants, 31, 65, 66
surgery, 7
surplus, 125
survival, xii, 19, 90, 112, 143, 154, 164, 165, 166, 184, 203
symmetry, x, 17, 28, 30, 49, 50, 123, 156
synergetics, 118

synthesis, x, 28, 29, 33, 35, 36, 39, 42, 44, 45, 47, 48, 60, 63, 72, 76, 77, 78, 79, 84, 88, 89, 91, 92, 117, 119, 120, 128, 132, 169, 197, 203

T

targets, 7
taxis, 120
teachers, 9, 11, 12, 19
teaching, 5, 8, 9, 10, 11, 12, 16, 17, 19, 20, 21, 25, 131, 132, 140
teaching process, 12
technical assistance, 93
television, 5, 14
tellurium, 92
temperature gradient, 31, 52
temporal, 119, 120, 127
tension, 17, 79, 105, 121, 123
tetroxide, 151, 152
textbooks, 19, 132
thermal decomposition, 47, 76
thermal properties, 196
thermal stability, 127
thermodynamic, 52, 128
thin film, 33, 163
thin films, 163
thinking, 9, 10, 11, 13, 16, 17, 22, 24, 199
Three Mile Island, 3
threshold, 155
time frame, 67
Titan, xi, 31, 44, 81, 83, 136, 137
titanium, 159
toluene, 43, 44
topographic, 98, 110, 201
toxins, 164
tradition, 6, 7, 20
transcription, 88, 89, 91
transcriptional, 89
transfer, 8, 11, 49, 75, 120
transfer RNA, 75
transformation, 116, 117, 119, 125
transformations, xii, 115, 118, 120, 123, 125, 128
transition, xii, 30, 56, 65, 115, 116, 117, 118, 119, 121, 123, 125, 128, 196
transition metal, 56
transition temperature, 65
transitions, 24, 117, 118, 123
translation, 88, 89, 91, 92, 128
translational, 89
transmembrane, 78
transparent, 153, 163, 175
transport, 68, 71, 88, 89, 108, 119
travel, 7, 17, 22
trees, 19, 87
tricarboxylic acid, 36
troposphere, 31
true belief, 9
tungsten, 42
two-dimensional, 91, 153
typology, 24

U

ultraviolet, xii, 29, 36, 37, 42, 62, 110, 143, 144, 153, 167, 169, 172, 178, 183, 191, 198
umbra, 98
uncertainty, 13, 115
undergraduate, xii, 131, 140
undergraduate education, 131
UNESCO, 23
uniform, 17
universality, 62
universe, 5, 12, 16, 17, 21, 82
urea, 36, 46, 47, 48
UV light, 62
UV radiation, 110
UV-radiation, 183

V

vacuole, 166, 174, 175
values, xii, 6, 52, 85, 98, 101, 103, 106, 109, 131, 132, 135, 136, 137, 139, 162, 196
van der Waals, 105, 111, 112
van der Waals forces, 105, 112
vapor, x, 28, 30, 53, 110, 139
variables, 53
variation, 8, 19, 70
velocity, 16, 108, 109
Venus, 4, 82, 137, 138, 139
versatility, 65, 74, 85
vesicle, 64, 65, 66, 67, 68, 70, 71, 73, 74, 75
vesicles, 44, 63, 64, 65, 66, 67, 68, 69, 70, 71, 73, 77, 78, 79
vessels, 11, 60
vibration, 156, 161
vibrational modes, 148, 151
virus, 60
viscosity, 88, 108
visible, 52, 99, 100, 101, 103, 106, 109, 133, 134, 136, 137, 139, 144, 150, 151, 153, 154, 156, 166, 169, 171, 172, 201
visualization, 139
voids, 138
volcanic activity, 61, 84

Vygotsky, 6, 25

W

waste products, 71, 88
water vapor, 30, 110
water-soluble, 33, 39
wavelengths, 38, 144, 146, 150, 151, 153, 164, 183, 187
weak interaction, 75
wind, 31, 99, 100
worldview, 21

X

XPS, 36, 37

Y

yield, x, 27, 29, 32, 35, 36, 40, 46, 49, 50, 51, 62, 134

Z

zirconium, 60